T0075857

The Cambridge Technical Series
General Editor: P. Abbott, B.A.

THE CHEMISTRY OF DYESTUFFS

THE CHEMISTRY OF DYESTUFFS

A MANUAL FOR STUDENTS OF CHEMISTRY AND DYEING

BY

M. FORT, M.Sc. (Leeds)

Late Lecturer in Dyeing in the Bradford Technical College

AND

L. L. LLOYD, Ph.D. (Bern)

Lecturer in Organic and Technical Chemistry in the Bradford Technical College

Cambridge:

at the University Press

1919

CAMBRIDGE
UNIVERSITY PRESS

University Printing House, Cambridge CB2 8BS, United Kingdom

Cambridge University Press is part of the University of Cambridge.

It furthers the University's mission by disseminating knowledge in the pursuit of education, learning and research at the highest international levels of excellence.

www.cambridge.org
Information on this title: www.cambridge.org/9781316606933

First edition 1917
Reprinted 1919
First paperback edition 2016

A catalogue record for this publication is available from the British Library

ISBN 978-1-316-60693-3 Paperback

PREFACE

THE recent opportunity offered to the dyestuff industry in this country may be expected to lead to a greater interest in the chemistry of dyestuffs and a greater demand for chemists possessing special knowledge of colouring matters. To the dyer, a scientific knowledge of dyestuffs is always very useful, and to the dyer's chemist it is indispensable.

It is necessary, however, that a thorough and systematic study of pure chemistry should precede any attempt to master a branch of applied chemistry, and we have presumed some such preparation by the student for whom this book is written. Although not specially intended to cover the syllabus for the City and Guilds of London Institute's examinations it provides the requisite knowledge in compact form.

The scheme of treatment adopted being on elementary text-book lines, we have made no attempt to give references to original papers and patents. This would have involved considerable extra cost, and moreover such references are usually neglected by students in a first study of the subject, in our opinion wisely so. In a second reading, however, works of reference should not be neglected, and the type of additional reading recommended in the first place is intermediate between text-book and descriptions of original research, namely, a résumé on some special section of subject matter. In addition, current literature must be consulted for new work and abstracts of new patents. It is only in this way that one can maintain contact with an industry developing with such extreme rapidity as the synthetic dyestuff industry has done hitherto. In any case, certain kinds of knowledge are most difficult to acquire, especially information as to the plant actually in use for large scale manufacture. So far as the limits of the book have allowed, we have endeavoured to assist the student in grasping the methods of large scale manufacture.

M. F.
L. L. L.

July 1916.

The second edition of this book follows so closely after the first that practically nothing has been altered. The activity of dyestuff manufacturers outside Germany has been directed mainly towards increasing the output of those dyestuffs most necessary to dependent industries. New names are being given to dyes previously manufactured in Germany, but no special introduction of new dyes has taken place since the first edition was printed.

M. F.
L. L. L.

June 1918.

CONTENTS

PART I

INTERMEDIATE COMPOUNDS

PART II

DYESTUFFS

BIBLIOGRAPHY OF WORKS OF REFERENCE

Handbuch der Organischen Chemie, F. Beilstein. Voss, Leipzig.

Die Fortschritte der Teerfarbenfabrikation und verwandter Industriezweige, P. Friedlaender, 8 vols. (1877—1911). Springer, Berlin.

Coal Tar and Ammonia, G. Lunge, 2 vols. (1909). Gurney and Jackson, London.

Technical Methods of Chemical Analysis, G. Lunge, trans. C. A. Keane, Volume II (1911). Gurney and Jackson, London.

Die Chemie des Steinkohlenteers, G. Schultz, 2 vols. (1900). Vieweg, Braunschweig.

Farbstoff-Tabellen, G. Schultz, 12 parts (1911—1914). Springer, Berlin.

Dictionary of Applied Chemistry, T. E. Thorpe, 5 vols. (1912–1913). Longmans, London.

JOURNALS.

Berichte der Deutschen Chemischen Gesellschaft (Berlin).

Journal of the Chemical Society (London).

Journal of the Society of Chemical Industry (London).

Journal of the Society of Dyers and Colourists (Bradford).

Journal of Industrial and Engineering Chemistry (American Chem. Soc., Easton, Pa.).

INDEX OF ABBREVIATIONS OF NAMES OF DYESTUFF MANUFACTURERS WHOSE PRODUCTS ARE DEALT WITH IN THIS VOLUME

British :—(R.H.) Read Holliday & Sons, Ltd., Huddersfield branch of British Dyes Ltd.

(Lev.) Levenstein Ltd., Manchester.

(B.A.) British Alizarin Co., London.

(Cl.) Clayton Aniline Co., Manchester. (This firm is now the property of the Society of Chemical Industry in Basle.)

French :—(P.) Société Anonyme des Matières Colorantes de St Denis, Paris (Poirrier & Dalsace).

(Mon.) Société Chimique des Usines du Rhône, Lyons (Gilliard, Monnet & Cartier).

American :—(Sch.) The Schoellkopf Aniline and Chemical Co., Buffalo, U.S.A.

Swiss :—(Gy.) J. R. Geigy, Basle.

(S.) Sandoz & Co., Basle.

(S.C.I.) Society of Chemical Industry, Basle.

German and Austrian :—(B.) Badische Anilin und Sodafabrik, Ludwigshafen (The Badische Co. Ltd.).

(By.) Farbenfabriken vorm. F. Bayer & Co., Elberfeld (The Bayer Co.).

(M.) Farbwerke vorm. Meister, Lucius & Brüning, Höchst (Meister, Lucius & Brüning Ltd.)

(C.) Leopold Cassella & Co., Frankfort.

German and Austrian, continued:—

(Ber.) Aktiengesellschaft für Anilinfabrikation, Berlin (The Berlin Aniline Co.).

(K.) Kalle & Co., Biebrich.

(G.E.) Chemische Fabriken Griesheim Elektron, Griesheim.

(W.-t.-M.) Chem. Fab. vorm. Weiler-ter-Meer, Uerdingen.

(J.) Carl Jäger, Barmen.

(Ki.) Kinzlberger & Co., Prague.

(L.) Farbwerk' Mühlheim (A. Leonhardt & Co.).

(W.) R. Wedekind & Co., Uerdingen.

The leading German firms were associated in two groups, the first including the Badische Anilin u. Sodafabrik, F. Bayer & Co., and the Aktiengesellschaft, Berlin. The second combination included the firms of Cassella, Meister &c. and Kalle. For the purpose of further strengthening the German industry, these two trusts and the firm of Weiler-t.-Meer formed a new syndicate in 1916, with a capital of over £11,000,000.

CHAPTER I

HISTORICAL INTRODUCTION TO THE CHEMISTRY OF THE DYESTUFFS

In 1856 Perkin obtained a colouring matter from crude aniline by oxidation and put it on the market as "Mauve[1]." Until that date the whole of the available colouring matters were of natural origin, chiefly vegetable, along with a few insect products like cochineal and lac dye, and a few mineral pigments, e.g., Prussian blue and chrome yellow. Since Perkin's discovery there has arisen a synthetic dyestuff industry developing with unequalled rapidity, until now the human taste for colour in all kinds of fabrics and tissues, in paints, varnishes, foodstuffs, etc., is almost entirely met by employment of synthetic products which did not previously exist in Nature, the materials and methods of scores of earlier generations being discarded wholesale, except in a few instances where natural dyestuffs are still able to compete with the synthetic products for special requirements. The old established natural dyestuffs have not been displaced without a struggle, and even

[1] This colouring matter was in use for postage stamps until the end of Queen Victoria's reign, long after it had been superseded for other purposes.

now one finds remnants of a diminishing prejudice against "aniline dyes" in the popular mind, but scientifically regarded the distinction between natural and artificial dyestuffs is an arbitrary one, relating only to their origin. Thus we now have Alizarin, an artificial dyestuff, replacing Madder, not, however, by providing merely an imitation, for Alizarin is the actual colouring matter contained in Madder, only in a pure form produced synthetically.

The economic changes wrought by this revolution in the colouring industries have been far-reaching; large areas of land have had to be turned to fresh uses, and on the other hand a growing importance has become attached to certain materials previously regarded as waste products. Although the industry of dyestuff manufacture is relatively speaking quite young, economic changes of considerable import have occurred, and when war broke out Germany was the seat of the industry, while Switzerland held a distant second place as a producer of synthetic dyestuffs, Great Britain coming third. The reasons for this country's failure to retain the lead given to the home of a new industry are still the subject of occasional debate, but it may be stated that, although the accident of birth of "Mauve" was apparently greatly in Great Britain's favour, Germany was in every way more fitted to develop an industry which necessarily had to rest more and more on an absolutely scientific basis. As this gradually came about, the development naturally proceeded most rapidly in Germany, where the influence of Liebig and others had given a dignity and importance to chemical science such as it obtained in no other country. In consequence an army of workers in this branch of technical science was more readily mobilised, as was also capital for working industrial enterprises having a chemical basis. Other causes, such as the patent laws and the restrictive measures against the use of alcohol—an absolute necessity in the early days of the synthetic dyestuff industry—have also operated in Great Britain's disfavour.

Unfortunately it has been treated by certain German scientists and statesmen as a fair occasion for self-congratulation and a proof of superior national genius that this great industry should have found its chief seat in Germany. In the middle of last century British genius and capital were being applied to colonial expansion and the development of mechanical inventions which have since brought changes even more sweeping than have followed from the discovery of Mauve—and be it added, immensely more profitable to Great Britain than the coal tar dyestuff industry has yet been to Germany. The outbreak of war in 1914 however cut off the German supply and brought up the question as to whether the time had not come to encourage British enterprise in this industry. It is greatly to be hoped that the attempt will be successful, which it can, only by receiving scientific aid.

Of the comparatively few British chemists who have given serious attention to this branch of technical science, a relatively large number, comparing favourably with those of any other country, have achieved international fame, and as the British industry comes to demand more there need be no doubt of their being found.

Previous to the discovery of Mauve, picric acid and rosolic acid had been produced artificially without finding commercial use as colouring matters, although since then both have been used as such. Perkin's discovery was made valuable by an almost contemporary event, i.e., Mansfield's success in separating benzene from coal tar, in which it had been discovered by Hofmann. The benzene was crude, and on nitration and reduction gave a crude aniline, which was the substance employed by Perkin.

Research at that time was at an empirical stage, and aniline was treated indiscriminately with all kinds of reagents. In 1859 Verguin obtained Magenta, Fuchsin or Rosaniline from crude aniline. Girard and de Laire in 1860 obtained a spirit-soluble blue by heating Magenta with aniline. In 1862 Nicholson sulphonated Aniline Blue and made it water-soluble, thus obtaining the first true acid dye, the Soluble Blues, which are still in extensive use, as is also Magenta. In 1863 came Hofmann Violets, soon to be superseded by the Methyl Violets. About this time the researches of Hofmann, Rosenstiehl, and E. and O. Fischer, went far towards explaining syntheses in the triphenylmethane series, comprising Magenta and its derivatives. The period of activity in discovery of Rosaniline derivatives lasted until about 1877. Meanwhile colouring matters of other groups were being discovered: in 1862 Phosphine, the first yellow dye, by Nicholson; in 1863 Safranine, by Perkin; in 1864 Bismarck Brown, the first azo dye, and Martins Yellow, both by Martins and Caro.

The formulation of Kekulé's theory of the benzene ring in 1865 gave great impetus to synthesis in the field of dyestuffs. Empirical methods gave way before scientific methods of research, and the new era of scientific progress was strikingly inaugurated by the synthesis of Alizarin in 1869 by Graebe and Liebermann. The next ten years were very productive of new discoveries, especially in the class of azo dyes, now by far the most numerous class of dyestuffs in use. A corresponding falling-off in the discovery of Rosaniline derivatives was experienced, and since then activity in the production of new colouring matters of this class has subsided until now it is almost quiescent. On the contrary new azo dyes are constantly being brought out, and patent literature is still actively concerned with the registration of many valuable additions to the class.

In 1880 Adolf von Baeyer accomplished the first synthesis of indigo. Artificial indigo did not however become a commercial product for many years, and its present success is one of the most noteworthy achievements of chemical science backed by industrial enterprise.

The first direct cotton dye, Congo Red, was discovered in 1884 by Böttiger, and in 1887 the first cotton dye to be developed by diazotisation *in situ* on the fibre was discovered by Green, and appeared on the market as Primuline. Meanwhile the new class of insoluble azo dyes, produced on the fibre by successive treatments of cotton with the requisite components, was introduced in 1880 with Para-nitraniline Red (Para Red) by the firm of Holliday.

In 1893 Vidal patented the first important sulphide dye, Vidal Black, and a period of great activity in the discovery of new dyes of this class began, which has lasted up to the present but now shows distinct signs of falling off, at any rate until a more intimate knowledge of the structure of these compounds is rendered available. The last stage of development is marked by the discovery of Indanthrene Blue by Bohn in 1901, which was the beginning of a new series of vat dyestuffs, since enriched by many syntheses of indigo derivatives. Synthetic indigo had meanwhile begun to seriously compete with the natural product, and from 1906 to 1914 the acreage under indigo cultivation in India fell from 420,000 to 150,000. The cutting off of German synthetic indigo by the European war again stimulated indigo planting, but there are no available data.

At present great activity prevails in the synthesis of vat dyestuffs, which possess a unique degree of fastness along with valuable tinctorial properties. The demand for greater fastness has been most successfully met, not only by the introduction of new vat dyestuffs, but by many and great improvements amongst the older classes of synthetic dyes, notably in mordant dyes admitting of improved methods of dyeing. The term "aniline dye" or "coal tar dye" now only conveys reproach where densest ignorance prevails, and as this is yet by no means uncommon, it is carefully ministered to in certain cases, as in the Oriental carpet trade, where the steady demand for native workmanship has not prevented, nor been at all impaired, by the unobtrusive replacement of vegetable dyes used since remote antiquity by the generally superior synthetic colouring-matters, products entirely of our own age.

In close conjunction with the progress of the synthetic dyestuff industry has been the development of other synthetic aromatic products derived from the raw materials found in coal tar. New drugs, artificial perfumes, explosives, disinfectants, and many important bodies which do not fall into any of these classes, as for example saccharine, have been discovered and are manufactured by the colour firms as side-products of the great coal tar dyestuff industry.

Additional reference may be made to the trade journals from the outbreak of war in August 1914 to the establishment of the Government scheme in 1915. Also to the following papers:

"The Artificial Colour Industry," F. M. Perkin, *Jour. Soc. Dyers*, p. 338, 1914.

"The Coal Tar Colour Industry of England," I. Singer, *Jour. Soc. Dyers*, p. 124, 1910.

"Tinctorial Chemistry: Ancient and Modern," R. Meldola, *Jour. Soc. Dyers*, p. 123, 1910.

"The Relative Progress of the Coal-Tar Industry in England and Germany," A. G. Green, British Assoc. 1901, *Jour. Soc. Dyers*, p. 285, 1901.

CHAPTER II

TAR DISTILLATION

Production of Tar

I. The most plentiful supply of tar is obtained during the decomposition of bituminous coals for the production of coal gas. The coal is heated in retorts at about 900° C. to 1000° C., at slightly reduced pressure, to remove the gas from the hot retorts as quickly as possible. The gas is cooled artificially in order to condense the substances of low boiling point. After cooling, it is necessary to remove tar that is suspended in a very finely divided condition in the gas. This is done by passing the gas under about 20 inches of water pressure through a "Tar Separator." The whole of the condensed portion is then allowed to mix. This separates into an aqueous upper layer and a lower layer of tar. The aqueous layer contains ammonia and small quantities of tarry substances in solution.

II. A large, increasing supply of tar is obtained from coke ovens. The coal is heated to a higher temperature than in the production of coal gas.

The tar is much thicker and contains a larger quantity of free carbon than coal gas tar.

III. A fairly large amount of tar is obtained by cooling the waste gases from blast furnaces, and this tar contains phenols of high molecular weight and of high germicidal value, with a large amount of free carbon.

IV. During the preparation of charcoal by the distillation of wood, there is obtained a distillate which separates into two layers; the upper aqueous layer is used for the manufacture of acetone, methyl alcohol, and acetic acid; the lower layer furnishes wood tar from which creosote (crude guaiacol) is obtained.

Different classes of tar are also obtained from the following processes: the distillation of bituminous shale; the manufacture of oil gas; the partial decomposition of oils by "cracking," the oils being delivered in thin streams into very hot retorts, the distillation of crude mineral oils, etc.

A different variety of coal tar is obtained by the distillation of coal at comparatively low temperature, viz., at about 400° C. to 450° C.

Composition of Tar.

The composition of tar varies with the source, and in the case of coal tar with the coal and temperature of distillation.

Fatty and aromatic hydrocarbons have been extracted from coal by means of solvents, but the amounts of such substances are usually very small. Paraffin hydrocarbons and hexahydrofluorene have been obtained.

The yield, as well as the relative amounts of the constituents of the tar, varies with the class of coal. The brown or young coals (lignite) give more fatty compounds and also more tar than the short flaming or older coals. The older or the less bituminous the coals, the smaller will be the amount of low boiling point constituents in the tar, and the higher will be the amount of anthracene and like compounds.

By the distillation of coal, for the manufacture of patent fuel, at low temperature, viz. 400° C. to 450° C., there is obtained a tar, which is fairly rich in low boiling hydrocarbons, a large percentage consisting of paraffin hydrocarbons. The so-called benzol from this tar, on account of the presence of other than aromatic hydrocarbons, is of no use for the manufacture of nitrobenzene.

In the distillation of caking coals, for the manufacture of coal gas, it is general to calculate upon a yield of 10 gallons of tar per ton of coal. This figure is only a rough guide, the quantity varying very much with the kind of coal.

It has been shown recently by Pictet and others that coal from Montrambert yields a mixture of hydrocarbons by extraction with boiling benzol. Among others hexahydrofluorene has been separated. By the distillation of coal in vacuum a tar is obtained, free from phenolic substances, containing a large proportion of basic compounds, but practically free from the aromatic bodies naphthalene, anthracene. This vacuum tar is decomposed by ordinary distillation with production of benzol, naphthalene and anthracene. From this it appears that the aromatic compounds in tar are pyrogenetic decomposition products.

At about 900° C. to 1000° C. there is a maximum yield of benzene compounds with a good illuminating gas; higher temperatures give more gas, but of less illuminating power, and the tar is richer in the more complex aromatic compounds.

Coal tar is a black viscous liquid, Sp. Gr. 1·1 to 1·2, and contains many aromatic compounds, the chief of which are now given.

Hydrocarbons: benzene, toluene, xylenes and other homologues, naphthalene and its homologues, acenaphthene, fluorene, anthracene, phenanthrene, pyrene, chrysene, etc.

Neutral compounds: carbon disulphide, thiophene and its homologues, carbazol, etc.

Bases: pyridine and its homologues, quinoline, isoquinoline, aniline, acridine, etc.

Phenolic substances: carbolic acid, cresylic acids, naphthols, etc.

Practically all of these compounds are employed in the colour industry, although many are not separated as such from tar, but may be made from a more easily obtainable constituent, e.g., aniline from benzene, naphthols from naphthalene.

Gas tar contains roughly about 2 per cent. benzol, 0·5 per cent. toluol, 0·6 per cent. phenol, 5 to 6 per cent. naphthalene and 0·6 per cent. anthracene.

On distillation the tar is divided, as described later, into four or five parts.

The **light oil** contains roughly 5 to 15 per cent. phenols, 1 to 3 per cent. basic substances, 0·1 per cent.

thiophene and homologues, 0·3 per cent. nitriles, 1·5 per cent. neutral bodies containing oxygen, the remainder being benzene and its homologues.

The **middle oil** contains roughly 40 per cent. naphthalene, 25 to 35 per cent. phenol and homologues, the remainder consisting chiefly of pyridine and quinoline bases.

The **heavy oil** contains mainly cresols, quinoline bases, naphthalene and its homologues.

The **anthracene oil** contains 2·5 to 3·5 per cent. anthracene, 2·5 to 3·5 per cent. phenanthrene, 1·5 to 2·5 per cent. carbazol; along with fluorene, pyrene, chrysene, and phenolic substances.

Distillation of Tar.

The tar as obtained from the gas works contains some aqueous ammoniacal liquor; this is partly separated by allowing to stand in tanks. If the tar is run directly into the stills (Fig. I, Appendix) in this condition it must be distilled very slowly at first in order to avoid sudden boiling over; this may happen with careless working, sometimes attaining the violence of an explosion. Consequently in most works the tar is heated up to about 150° C. before running into the stills, the latter being usually hot from the last charge. Several devices are employed to secure this preliminary heating. The tar may be heated (dehydrated) by utilising the waste gases from the stills before passing up the chimney stack. This is sometimes done in large boilers or by means of a shallow tubular still; the products that pass off are condensed and collected. Water, carbon disulphide, pyridine, benzene, etc., are thus obtained. The water layer is then separated and the oil treated along with the light oils. The dehydrated tar is then run into the stills and directly distilled. The distillate is collected in fractions according to the specific gravity, or in some cases the fractions may be divided according to the temperature at which they distil.

The following table shows the approximate specific gravity and boiling points of the different fractions:

Fractions	Specific Gravity	Boiling Point	Average Yield
I. First Runnings or Light oil	0·9 to 0·95	up to 170° C.	2 to 6 °/ₒ
II. Second Runnings or Middle oil...	0·95 to 1·01	up to 230° C.	10 to 12 °/ₒ
III. Third Runnings or Creosote oil or Heavy oil	1·01 to 1·05	up to 280° C.	8 to 10 °/ₒ
IV. Anthracene oil ...	1·05 to 1·1+	to close of distillation	16 to 18 °/ₒ
V. Pitch	—	still content	about 50 °/ₒ

During the distillation of the light oil it is necessary to cool the distillation products in a water-cooled worm, but the higher boiling fractions are condensed without water-cooling on account of the danger of sealing up the worm by deposition of solid naphthalene. As soon as the creosote oil fraction has been distilled from the tar the distillation is completed by means of superheated steam. When distillation is complete the tar is run into a cooler, which consists in most cases of a large tank or boiler, in order to cool it below its firing point.

The tar is then run out and exposed to air in large shallow pits to solidify.

Continuous distillation of Tar. Several types of continuous distillation plant are in use. One of the best forms of these installations is briefly as follows. The tar is gradually warmed up by using it in the first place to run around the condensing worm of the first still, from which it flows to the other tanks through which the condensing worms pass. In this manner the tar is partly distilled, the volatile portions being collected before it enters the first still. The tar, as a rule, passes through three stills, the first and second being directly heated, and the third distillation conducted by means of steam. The tar flows continuously through the three stills, in shallow layers, in a circuitous course, and finally to a cooling chamber or in some cases directly to the solidifying pits. In this manner four fractions are collected which are similar to those stated above.

Treatment of Tar distillates.

The separate fractions as obtained by the distillation are now further treated in order to produce purer compounds as used in the chemical trades. The light oil is well mixed with caustic soda of 15° Tw. The acidic substances (phenols) are dissolved, the upper layer is run off, washed with water to remove alkali and then treated with sulphuric acid (1 part H_2SO_4 and 2 parts water). The sulphuric acid forms salts with bases, these remain in solution. The oil that collects above is well washed and then fractionated (Fig. II, Appendix), a Savalle column, a dephlegmator, or in some cases a combination of these, being used in order to obtain, by one distillation, practically pure benzene, toluene and commercial xylene.

Intermediate fractions are also obtained, these are added to fresh quantities of treated light oils, or are again redistilled.

The "**dephlegmator**" is the class of still-head in which the condensed liquid is caused, by means of suitable obstructions, to collect in shallow pools, through which the ascending vapour has to force its way; very good contact is thus obtained between vapour and liquid at definite intervals. The excess of liquid is carried back from pool to pool, and finally to the still by suitable trapped reflux tubes or intermittent syphons.

The **Savalle column** contains a dephlegmator at the lower portion, and then a surface, or multitubular, condenser, provided with a water supply, so regulated that its temperature is about that of the boiling point of the liquid required; the liquid condensed in the regulated temperature still-head returns to the dephlegmator, and the purified vapour passes on to the cold condenser.

The benzene and toluene so obtained still contain sulphur compounds, viz. thiophene and thiotolenes; these may be removed by extracting three or four times with small quantities of concentrated sulphuric acid. Thiophene and its homologues are sulphonated in the cold, whereas benzene is only slightly affected. For most purposes the thiophene is seldom removed from the benzene. The portions boiling above the boiling point of the xylenes furnish **solvent naphtha** or cumol (commercial), which is used in the rubber industry.

The middle oil on standing deposits dark brown crystals of crude naphthalene; these are separated by filter pressing or by a centrifugal machine. The oil so obtained is treated with caustic soda of 15° Tw. at about 40° C. and the lower caustic layer run off.

Stronger caustic soda solution may be used for the extraction of the phenols; such a solution is, however, a better solvent for the neutral oils, consequently the phenols obtained will be less valuable. The upper layer on cooling deposits a further quantity of crude naphthalene. The alkali extracted oil is, as a rule, added to the creosote oil, or in some cases is treated with sulphuric acid exactly the same as with the light oils. The neutral oil is then added to the creosote oil.

The caustic soda extracts of the first and second distillates are mixed together and treated for the separation of carbolic acid and cresols.

In some cases the alkali extract is just acidified with sulphuric acid, the phenols collect as an upper oily layer; the latter is separated, washed free from acid and distilled, a fractionating column (Savalle) being used to aid in the separation of the carbolic acid from the cresols. By using sulphuric acid for "cracking out" the phenols, there is a loss of about two to three per cent. of phenol which remains dissolved in the aqueous liquor. On account of this loss it is more general to separate the phenolic substances from the alkaline extracts by means of carbon dioxide. The furnace gases are employed for this purpose. They are first washed to remove sulphur dioxide, and then passed through the alkaline extract until the phenols are precipitated. The upper oil layer is washed and then distilled, and the lower aqueous solution of sodium carbonate containing a little phenol, etc. is recausticised with milk of lime and again used for extraction purposes.

The distillates from the phenols are collected and classified according to the melting points; 60° phenol means that the phenol has a mean melting point of

60° F. By repeatedly recrystallising it is possible to obtain a carbolic acid of melting point 38° C., but further purification, to obtain pure phenol, is impossible by simple crystallisation, owing to the presence of orthocresol, and, since separation cannot be effected by distillation, a method of purification, based on the formation of a crystallisable hydrate of phenol, is employed. The mixture is churned with warm water and allowed to cool, a phenol hydrate, having the formula $C_6H_5OH . H_2O$, crystallises out, whereas orthocresol forms a liquid hydrate. The crystals are separated and the phenol obtained by distillation. The phenol so obtained turns red on exposure to air and light, but is sufficiently pure for manufacturing purposes. The red colour is probably due to some sulphur compound which may be present in the original tar or produced by the action of too strong sulphuric acid in precipitation, etc. Since phenol "cracked out" by furnace gases also turns red it is most probably due to some compound originally present in the tar.

The cresols are mainly used for the manufacture of disinfectants, such as **Lysol**, **Jeyes' Fluid**, **Cresolin.**

In this connection it has been observed that the higher the boiling point of the phenol homologues the higher is the antiseptic value. Consequently the phenols from blast furnace gases are particularly important for the manufacture of these compounds.

The crude naphthalene as obtained above is melted and extracted with caustic soda to remove phenols, then with diluted sulphuric acid as above to remove basic compounds. The partly treated naphthalene is then hot pressed, similarly to the separation of stearin for candle manufacture. The homologues of naphthalene, etc., melt at a lower temperature and are expressed from the mixture. The now practically pure naphthalene is redistilled or sublimed in a hooded iron pot. It is used very extensively in the colour industry, also as a deodorant.

The basic compounds are separated from the sulphuric acid extracts by means of caustic soda or more generally by ammonia. The bases, pyridine and its homologues, are separated and distilled. The pyridine is used for "denaturing" alcohol and as a solvent for the purification of anthracene.

Pyridine and Quinoline bases. Pyridine and its homologues are at present only used as solvents for the purification of organic compounds. Dyestuffs are obtainable from these compounds, but they have not yet found commercial application.

Crude quinoline bases are obtained from coal tar and bone oil, and the portion boiling between 235° C. and 245° C. is employed to a small extent in the production of **Quinoline Red** and **Quinoline Blue.**

The synthetic compound **quinaldine** 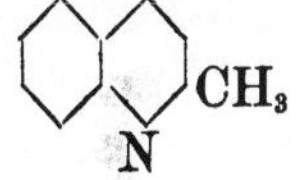

(B.P. 243° C.) is obtained by boiling aniline, hydrochloric acid, and para-acetaldehyde under a reflux condenser. The reaction mixture is made alkaline and the base distilled.

It is employed in the production of **Quinoline Yellow.**

The creosote oil contains higher phenols, naphthols, naphthalene homologues, etc.; it is used to a large extent for creosoting timber. For this purpose the wood is packed in tanks, the air pumped out, and when a good vacuum is obtained the oil is run in. In this manner the wood is well impregnated with the oil. The oil also finds application as a fuel. It is used for the manufacture of lamp-black, and in the waterproofing of felt. It is also added to pitch to soften it.

The anthracene oil contains anthracene, methyl-anthracene, naphthalene, phenanthrene, acenaphthene, fluorene, chrysene, pyrene, acridine, some naphthols and some quinoline bases. On allowing to stand for some time it separates into a crystalline and a liquid portion. This is filtered by filter-presses or by hydro-extractors (centrifugated). The liquid portion so obtained, which no longer crystallises on standing, is termed "**green oil**"; it is used for the preparation of cheap lubricants, and is also added to pitch for softening purposes.

The solid portion is expressed warm, which leaves a cake containing about 25 to 35 per cent. anthracene. The warm expressed oily portion is treated again after it has crystallised.

The crude 25 to 35 per cent. anthracene is purified by means of solvents until about 80 per cent. anthracene is obtained. For this purpose the following solvents may be used: solvent naphtha, crude pyridine, liquid ammonia, liquid sulphur dioxide, etc. (Fig. IX, Appendix.) When solvent naphtha is used, the extract is separated hot, leaving the anthracene of about 70 to 80 per cent. purity.

Crude pyridine gives a better separation than naphtha, the anthracene being only sparingly soluble is thus obtained as a greenish solid of about 80 per cent. purity.

The anthracene in this form is distilled in steam, and the distillate quickly chilled in order to obtain the substance in a finely divided form, in which state it is oxidised by sulphuric acid and sodium bichromate into anthraquinone.

Phenanthrene is not so readily oxidised as anthracene and may be extracted by solvents from the crude anthraquinone. The anthraquinone may be further purified by heating with concentrated sulphuric acid to sulphonate impurities, which may be then removed with alkali washing, after which 95 to 98 per cent. anthraquinone is obtained.

Phenanthrene, acenaphthene, fluorene, and carbazole are obtained from the naphtha and pyridine extracts. The commercial importance of these bodies has greatly increased of late years.

Carbazole is obtained by heating the crude or treated anthracene distillate with strong caustic potash; the resulting melt is allowed to cool and the upper layer stripped from the lower one of potassium carbazole, which on treating with water gives carbazole and caustic potash solution. Carbazole may be nitrated, sulphonated, halogenated, etc., the products being employed for the production of dyestuffs, e.g., **Carbazol Yellow** from diazotised diamino-carbazole and salicylic acid, **Hydron Blue** by polysulphide melt of carbazole indophenol.

The pitch, which remains in the stills, is run into cooling vessels, then into shallow pits, and is then broken up. The pitch is used for road material, asphalt, for the manufacture of briquettes, etc.

For some purposes the pitch is required free from suspended carbon; this is obtained by dissolving in twice its volume of hot high-boiling point naphtha, allowing the carbon to deposit, running off the clear hot liquid portion and then distilling off the solvent. Soft pitches, e.g., for modern road making requirements, are made by working up crude pitch with creosote and waste anthracene distillates.

Further information may be obtained from Lunge's *Coal Tar and Ammonia* (2 vols.), and "Tar Distillation" by H. P. Hird, *Jour. Soc. Dyers*, April, 1916.

PART I

METHODS FOR THE PREPARATION OF INTERMEDIATE COMPOUNDS

CHAPTER III

NITRO COMPOUNDS

THE ordinary chemical reactions that are employed for the preparation of the intermediate compounds may be classified under nitration, sulphonation, chlorination, amidation, alkylation, arylation, acylation, hydroxylation, reduction, oxidation, condensation, and diazotisation.

Nitration. The introduction of nitro groups into aromatic compounds is one of the most important reactions, and is capable of application to a large variety of compounds. There are many methods of nitration, the selection of the method depending upon the nature of the compound to be nitrated. For example, phenol is more easily nitrated than benzene. As nitrating agents one employs nitric acid, potassium nitrate and sulphuric acid, nitric acid and sulphuric acid. The temperature at which nitration is carried out is, as a rule, very important, since the higher the temperature, the more readily will tar be formed, and, generally, the higher will be the degree of nitration.

In many cases the sulphuric acid is added as a dehydrating agent in order to prevent the dilution of the nitric acid, so that the whole of the nitric acid is utilised for nitration purposes. In other cases it is desirable that sulphonation should precede nitration, to facilitate the latter, the sulphonic acid group being replaced by the nitro group. Of great importance is the nature and position of groups present in the aromatic compound.

In some cases, owing to the oxidising action of nitric acid, it is necessary to shield certain groups, as in the preparation of nitranilines from acetanilide.

Nitration of benzene. Nitrobenzene $C_6H_5NO_2$ (B.P. 205° C.) is manufactured from commercial pure benzene (thiophene present) in iron vessels by means of nitric and sulphuric acid. Sulphuric acid is added to combine with the water originally present in the nitric acid and also that formed during nitration. The quantity of sulphuric acid should be sufficient to produce the hydrate $H_2SO_4 . 2H_2O$. In this manner the whole of the nitric acid added is utilised for nitration. The acid mixture is slowly run into the benzene, and well agitated by means of iron stirring gear. The vessel is provided with a leaden or earthenware cooling worm, through which cold water may be run if the temperature of the contents of the vessel rises too high. During the beginning of the nitration the temperature is seldom allowed to exceed 40° C., but towards the finish the temperature may be raised to about 75° C. After the necessary time for nitration, the contents are allowed to stand until separated and the lower acid layer run off. This is concentrated, and again employed for the same purpose.

The nitrobenzene that collects above the acid liquor is separated and well washed to remove acid, and is then distilled, either directly or by steam, "pure" nitrobenzene or **Mirbane oil.** It may be obtained pure enough to use as a cheap scent by repeated steam distillation. It has an odour resembling bitter almond oil. The "pure" oil is used in the manufacture of aniline for the production of **aniline black**; also for the preparation of azobenzol, benzidine, **Induline**, etc. Nitrobenzene containing nitrotoluene is termed "nitrobenzene for red," as it is used in the manufacture of **Magenta**, and also **Phosphine.** Nitrobenzene is largely used as a solvent when reactions are to be conducted at fairly high temperatures. It may also be employed as an oxidising agent.

Nitrotoluenes are obtained in a similar manner to the manufacture of nitrobenzene. The crude nitrotoluene should not be distilled with direct fire on account of the liability to explosion. In this case there are two bodies, ortho- and para-nitrotoluene, produced. The two isomers are separated by fractional distillation in vacuum, the para compound being further purified by crystallisation, **o-nitrotoluol** B.P. 223° C., **p-nitrotoluol** M.P. 54° C. They are employed for the production of the corresponding toluidines, etc. In some cases the two nitro compounds are not separated, but are converted into mixed toluidines, the toluidines being more readily separated from one another than the nitro compounds.

Toluene is more easily nitrated than benzene.

Meta-Dinitrobenzene (benzene ring with NO_2 at the side and NO_2 below) (M.P. 90° C.) is obtained by a similar but stronger nitration in iron vessels; the nitrating acid is added in two portions, and the temperature is finally raised to about 100° C. The lower acid layer is run off, and the dinitrobenzene washed well with cold water and then with hot water, to prevent it from solidifying during washing. The latter washing will contain some dinitrobenzene and is used after cooling for the cold washing of a fresh batch. The dinitrobenzene is run into iron trays about 2 to 4 inches deep and allowed to solidify.

Small quantities of ortho- and para-dinitrobenzenes are also produced.

Dinitrotoluene is obtained similarly, the chief product being the 2.4-dinitrotoluene.

Nitrochloro-benzenes are obtained similarly to the nitrotoluenes, and are separated by similar methods. **Ortho-nitrochloro-benzene** B.P. 246° C., M.P. 32° C. **Para-nitrochloro-benzene** B.P. 239° C., M.P. 83° C.

Dinitrochloro-benzene is obtained by further nitration of mono-chloro-nitrobenzene. The main product is **2.4-Dinitrochloro-benzene** M.P. 50° C., B.P. 315° C. This compound is chiefly used for the production of dinitrophenol.

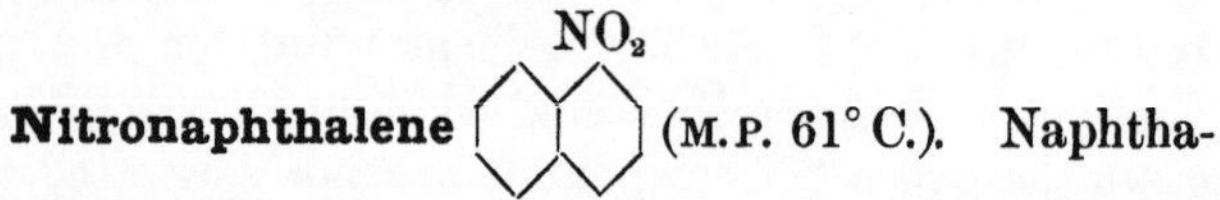

Nitronaphthalene (M.P. 61° C.). Naphthalene, in a fine state of division, is well mixed with nitrating acid, the temperature during the action being maintained at 45° to 50° C. After cooling, the cake of α-nitronaphthalene is washed similarly to the purification of meta-dinitrobenzene.

On further nitration of nitronaphthalene a mixture of 1.5- and 1.8-dinitronaphthalenes is obtained. This consists roughly of one part of the 1.5- and two parts of the 1.8-dinitronaphthalenes. For many purposes the crude dinitronaphthalene is directly employed. These isomers may be separated from each other by means of cold pyridine, the 1.5-derivative being less soluble (1 : 125) than the 1.8-isomer (1 : 10).

1.5-Dinitronaphthalene M.P. 216° C.
1.8- „ „ M.P. 170° C.

After the introduction of two nitro groups a third may be introduced in the β-position and thus by the further nitration of naphthalene, tri- and tetra-nitronaphthalenes may be obtained.

Nitration of phenol. By the careful nitration of phenol with dilute nitric acid a mixture of ortho- and para-nitrophenol is obtained. The temperature during nitration is not allowed to exceed 25° C., in order to prevent formation of tar. The dilute nitric acid is obtained by the action of dilute sulphuric acid on sodium or potassium nitrate. The ortho derivative is then separated from the para by steam distillation.

Ortho-nitrophenol (M.P. 44° C.) is a light yellow body used in the manufacture of ortho-nitroanisol for the preparation of anisidine and dianisidine.

Para-nitrophenol (M.P. 114° C.) is non-volatile in steam, and is colourless.

The dinitrophenols may be obtained by nitration of phenol, but as a rule the 2.4-dinitrophenol is obtained by boiling 1-chlor-2.4-dinitrobenzene with a strong solution of sodium carbonate under a reflux condenser. A simple and effective type of condenser is an upright iron pipe with cold water running down the outside. The dinitrophenol is then precipitated by acidifying with sulphuric acid. It is a yellow compound, M.P. 114° C., and is employed for the preparation of sulphide blacks.

Picric acid $C_6H_2(OH)(NO_2)_3$ [OH at 1; O_2N, NO_2 at 2, 6; NO_2 at 4] (M.P. 122·5° C.) or symmetrical trinitrophenol is manufactured by sulphonating phenol at 100° C., generally in earthenware vessels. To the sulphonated phenol nitric acid is added with stirring, the picric acid is allowed to crystallise out, the acid liquor run off, the crystals washed with cold water and then recrystallised from hot water. Owing to the explosive nature of picric acid and its salts, especially in presence of traces of acids, or metallic oxides, the purification must be carefully attended to, explosions having been brought about by traces of such substances as whitewash, lime, etc.

Picramic acid [OH at 1; O_2N at 2, NH_2 at 6; NO_2 at 4] (M.P. 165° C.) is obtained by reducing a warm aqueous solution of picric acid with zinc dust and ammonia; when the solution remains distinctly alkaline the excess of ammonia is removed by boiling, the solution filtered and concentrated, and the picramic acid precipitated by addition of acetic acid. It is a red crystalline compound.

CHAPTER IV

AMIDO COMPOUNDS

Amidation. By amidation is meant the preparation of amido compounds by the introduction of the NH_2 group. The most important method for the production of amido compounds is by reduction of nitro compounds. As reducing agents, iron in presence of hydrochloric or acetic acid is the most important, and to a smaller extent zinc, tin, sodium sulphide, sulphurous acid and its salts, sodium hydrosulphite, caustic soda and glucose, etc. Of great importance is the nature of the reduction, whether it be in acid, neutral, or alkaline solution. For example, nitrobenzene gives on reduction in acid solution aniline $C_6H_5NH_2$, in neutral solution phenylhydroxylamine C_6H_5NHOH, and in alkaline solution hydrazobenzene $C_6H_5 . NH . NHC_6H_5$.

NH_2

Aniline (benzene ring) (M.P. 8° C., B.P. 182° C.) is manufactured in iron vessels by reducing nitrobenzene with scrap iron and a small amount of hydrochloric acid. A large horizontal boiler is fitted with a propeller or shovel stirring gear and also a hollow axle constructed for steam distillation. Iron filings, nitrobenzene, water, and about 3 per cent. of hydrochloric acid on the weight of nitrobenzene are introduced into the vessel. The contents are well stirred, and steam passed into the vessel. Water, nitrobenzene, and aniline distil; the distillate is condensed and the oil returned to the reduction vessel until the nitrobenzene is completely reduced. This is observed from the colour, smell, and specific gravity of the oily distillate. The condensed and separated water, containing 3 per cent. of aniline, is run back to the steam boiler. The aniline may be rectified or again steam distilled to further purify it. When pure it is a colourless, highly refractive liquid. The commercial aniline may contain about 2 per cent. of nitrobenzene, and on keeping changes colour to yellow or red, this change being probably due to a trace of amido-thiophene. Aniline for the production of aniline black should be tested for the toluidines and nitrobenzene, or reddish azine dyes will be formed along with the black.

The so-called "aniline for red" is a mixture of aniline with ortho- and para-toluidines, obtained from the mixed nitrotoluenes, and it is employed for the manufacture of **Magenta.**

The reactions that take place when nitrobenzene is reduced with iron may be represented as follows:

I. $Fe + 2HCl = FeCl_2 + 2H.$

II. $6FeCl_2 + C_6H_5NO_2 + 4H_2O$
$= 4FeCl_3 + C_6H_5NH_2 + Fe_2(OH)_6.$

III. $2FeCl_3 + Fe = 3FeCl_2.$

From these equations it is obvious why iron in presence of a small amount of acid is capable of reducing a large amount of nitro compound.

The full equation may therefore be expressed:

$$C_6H_5NO_2 + 2Fe + 4H_2O = C_6H_5NH_2 + Fe_2(OH)_6.$$

Aniline, being a basic compound, is capable of producing salts; one of the most important is "**Aniline Salt**" or aniline hydrochloride $C_6H_5NH_2 . HCl$. It is obtained by mixing aniline with concentrated hydrochloric acid in molecular proportions and then allowing the aniline salt to crystallise out. Aniline sulphate is prepared similarly.

Other amines prepared in a similar manner to aniline are ortho- and para-toluidine, the xylidines, and α-naphthylamine. In some cases the nitrotoluenes are not separated, but are converted into toluidines by direct reduction of the crude nitrotoluene; the crude product consists of about 32 to 38 per cent. of para-, 60 to 65 per cent. ortho-, and about 2 per cent. of meta-toluidines. By strongly cooling, the para-isomer crystallises out, and may be filtered, or, better, the mixture is treated with sodium phosphate (Na_2HPO_4) solution, with which the para-toluidine forms an insoluble compound; but probably the cheapest method is that of churning and cooling with water or with ice. The para-toluidine forms a crystalline hydrate which may be easily removed and purified. Para-toluidine may be detected in the presence of aniline and o-toluidine by addition of ferric chloride to a hydrochloric acid solution of the mixture. With the para compound a Bordeaux red solution is obtained. If o-toluidine and aniline are present a greenish blue precipitate is formed, which on filtration gives a red filtrate.

The xylidines are usually prepared mixed from unseparated nitroxylenes; the meta-xylidine is the most important.

In the case of naphthylamine ordinary steam distillation is not sufficient to separate it from the sludge. Superheated steam may, however, be used; or, preferably, extraction with solvent naphtha is employed. All these amines are in use for the manufacture of azo dyes, besides the similar use of a large number of their derivatives.

Derivatives of Aniline. There are two classes of derivatives, those in which the substitution takes place in the ring and the amido substitution derivatives. Of the latter class, the most important are mono-methyl-aniline, dimethyl- and diethyl-aniline, diphenylamine, and benzylaniline.

The ring-substituted derivatives of aniline are very important, of which the following may be taken as typical examples: ortho-, meta-, and para-nitranilines, chloroanilines.

The introduction of the nitro group into aniline gives compounds solid at the ordinary temperature. These compounds are far less basic than aniline so that their salts are hydrolysed by water with separation of the free base. Aniline hydrochloride on the other hand is stable even in very dilute aqueous solution.

On account of the oxidising action of nitric acid upon aniline it is impossible to nitrate directly. The amino group is therefore shielded by acetylation.

Acetanilide $C_6H_5NH\,COCH_3$ is obtained by boiling aniline and glacial acetic acid in an enamelled iron

$$C_6H_5NH_2 . CH_3COOH \rightarrow H_2O + C_6H_5NHCOCH_3$$

vessel, fitted with a reflux condenser, until the boiling point of the mixture reaches 250° to 270° C. The product is run into trays where it sets on cooling. It is ground to powder in a mill and nitrated directly. The nitration is carried out in an iron vessel similarly to the manufacture of nitrobenzene. The finely powdered acetanilide is well mixed with some concentrated sulphuric acid; into this mixture the theoretical amount of nitrating acid (calculated upon the nitric acid) is gradually added. The contents of the vessel are thoroughly agitated and the temperature kept between 10° and 15° C. The reaction product is run into cold water in a lead lined vessel and thoroughly mixed with a revolving paddle agitator. The precipitated **para-nitracetanilide** (benzene ring: $NHCOCH_3$ above, NO_2 below) is filter-pressed, washed with water and hydrolysed. There is also obtained about 4 per cent. of ortho-nitracetanilide; this may be removed by chloroform or water.

The para-nitracetanilide is hydrolysed by boiling with 25 per cent. sulphuric acid. The **para-nitraniline** (benzene ring: NH_2 above, NO_2 below) is precipitated in a finely divided state by running into cold water with agitation, filtered, and dried. It is applied for the production of **Para Red** and in the manufacture of azo dyes. It is a yellow compound, non-volatile with steam, and melts at 147 °C.

Pure para-nitraniline is most readily obtained by nitrating benzilidene aniline (benzene ring)$N=CH$(benzene ring), the nitro derivative being then hydrolysed with dilute caustic potash.

Nitracetanilide on reduction gives para-amidoacetanilide, which is largely used for the production of azo dyes.

In order to obtain a larger yield of **o-nitraniline** (M.P. 71·5° C.), acetanilide is treated with sulphuric acid at 50° C., followed by addition of nitrating acid, keeping the temperature at 40° to 50° C. The product is precipitated with water, deacetylated by steam, cooled to 50° C., and poured on ice. The o-nitraniline is precipitated, the para compound remaining in solution. A yield of 25 per cent. of the ortho compound may be obtained.

Meta-nitraniline ($C_6H_4(NH_2)NO_2$, benzene ring with NH_2 and NO_2) (M.P. 110° C.). This is obtained by reduction of one nitro group of m-dinitrobenzene. Such partial reduction is effected by prolonged action of cold sodium sulphide solution. A similar method is used for reducing nitro groups in azo dyes without the azo group also suffering reduction.

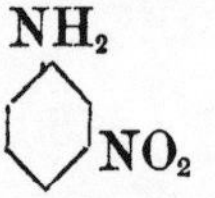

Aniline Sulphonic Acids. There are three isomerides of which the para compound, usually called sulphanilic acid, is the most important, and secondly the meta or metanilic acid. These acids are used commercially in large quantities for the manufacture of azo dyes, etc.; the ortho acid does not yet find application commercially.

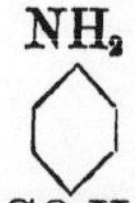

The para-aniline sulphonic or **sulphanilic acid** (benzene ring with NH_2 and SO_3H) is manufactured by the "baking process." Molecular quantities of aniline and sulphuric acid are mixed together with stirring, aniline sulphate is formed; this solid substance is heated in shallow layers upon trays at 180° to 220° C. until the product is completely soluble in dilute caustic soda. The product is dissolved in water, filtered from carbon, and allowed to crystallise.

Metanilic acid $C_6H_4(NH_2)(SO_3H)$ is obtained by sulphonating nitrobenzene with fuming sulphuric acid, running the product into water and reducing by means of iron filings.

Aniline Carboxylic Acids. By the nitration of benzoic acid a mixture of three isomers is obtained consisting of about 80 per cent. meta-nitrobenzoic, 15 per cent. para-nitrobenzoic and 5 per cent. of ortho-nitrobenzoic acid.

Ortho-nitrobenzoic or Anthranilic acid $C_6H_4(NH_2)(COOH)$ (M.P. 145° C.) is much employed in the dyestuff industry. It is obtained in several stages from naphthalene, the first being oxidation with

$$C_{10}H_8 + \text{oxygen} \rightarrow C_6H_4(COOH)_2 + H_2O$$

sulphuric acid in presence of mercury sulphate; at 250° to 270° C. sulphur dioxide, carbon dioxide, and water are given off, at 300° C. phthalic and its anhydride distil. The mercury sulphate remains behind and is used for the next charge. **Phthalic acid** $C_6H_4(COOH)_2$ (M.P. 213° C.) readily loses water on heating forming **phthalic anhydride** (M.P. 128° C., B.P. 284° C.). It is used for the manufacture of **Fluorescein, Rhodamine,** etc.

$$C_6H_4(COOH)_2 \rightarrow H_2O + C_6H_4{<}^{CO}_{CO}{>}O$$

Phthalimide (M.P. 238° C.) is obtained by heating phthalic anhydride and ammonium carbonate at 225° C. until the mass, which at first liquefies, again solidifies. After cooling it is dissolved in water and recrystallised.

$$C_6H_4\left<\begin{matrix}CO\\CO\end{matrix}\right>O + NH_3 \rightarrow C_6H_4\left<\begin{matrix}CO\\CO\end{matrix}\right>NH + H_2O$$

The phthalimide is dissolved in cold caustic soda solution and treated with the theoretical quantity of sodium hypochlorite (Hoffmann's Reaction), the solution is then quickly heated to 80° C. at which temperature the reaction rapidly takes place. The solution is neutralised with sulphuric acid and the **anthranilic acid** precipitated by addition of acetic acid. It is used in the synthesis of indigo and for preparing thiosalicylic acid, which is used in the synthesis of certain indigoid vat dyes.

$$C_6H_4\left<\begin{matrix}CO\\CO\end{matrix}\right>NH + O + NaOH \rightarrow C_6H_4\left<\begin{matrix}COONa\\NH_2\end{matrix}\right. + CO_2$$

N-substituted aniline derivatives. The alkyl and aryl derivatives of aniline are extensively used in the manufacture of the basic dyestuffs.

Monomethylaniline $C_6H_5NH.CH_3$ is obtained by heating aniline, aniline hydrochloride and methyl alcohol (free from acetone) under pressure in an autoclave to 200° C. The bases are separated by milk of lime followed by steam distillation. The oil separated from the distillate is a mixture of aniline, and mono- and dimethylaniline. This oil is acetylated, as in the preparation of acetanilide, and then distilled. The dimethylaniline distils off, the residue is hydrolysed by boiling with dilute sulphuric acid, then made alkaline with milk of lime, and distilled. Monomethylaniline is a colourless oil (B.P. 191° C.).

Monoethylaniline $C_6H_5NHC_2H_5$ is obtained similarly (B.P. 204° C.).

Benzylaniline $C_6H_5NH.CH_2C_6H_5$ is obtained in the same way except that benzyl chloride is substituted for the alcohol, and the reaction temperature is lower, viz., 160° to 170° C.; M.P. 31° to 33° C.

Dimethylaniline $C_6H_5N(CH_3)_2$ (B.P. 192° C.). This compound is manufactured by heating aniline, aniline hydrochloride, and methyl alcohol in an autoclave at 200° to 230° C. during ten hours (Fig. V, Appendix). The reaction product is made alkaline with milk of lime and the bases separated by steam distillation. The oil may then be rectified by distillation. As impurity there is generally a little monomethylaniline. Dimethylaniline is a colourless oil, and is largely used in the colour industry.

Diethylaniline $C_6H_5N(C_2H_5)_2$ (B.P. 213° C.) is obtained similarly, it is a colourless oil.

Diphenylamine $C_6H_5NHC_6H_5$ (M.P. 54° C., B.P. 310° C.) is obtained by heating aniline and aniline hydrochloride in an autoclave at 220° C. (Fig. III, Appendix). The reaction product is dissolved in boiling concentrated hydrochloric acid, poured into a large quantity of water, and the base, which separates out, is redistilled. A blue colour is obtained with traces of nitrates, nitrites or chlorates in strong sulphuric acid.

3—2

Naphthylamines. **α-naphthylamine** [structure: naphthalene ring with NH_2 in alpha position] (M.P. 50° C., B.P. 300° C.), see p. 29.

β-naphthylamine [structure: naphthalene ring with NH_2 in beta position] (M.P. 112° C., B.P. 294° C.). Since by the nitration of naphthalene only the alpha position is substituted, the β-naphthylamine must be obtained from naphthalene β-sulphonic acid or from β-naphthol. One molecule each of β-naphthol and of ammonium sulphite, and one and three-quarter molecules of ammonia (25 per cent. solution) are heated in an autoclave at 150° C. during about ten hours. The crude product is treated with caustic soda to remove unaltered β-naphthol, then dissolved in hydrochloric acid, filtered from the insoluble β-dinaphthylamine, and separated from the hydrochloric acid solution by milk of lime or soda. When pure it is colourless and almost odourless.

The addition of ammonium sulphite greatly reduces the temperature at which the reaction takes place. The reaction is probably as follows: a sulphurous acid ester is first obtained:

$$C_{10}H_7OH + (NH_4)_2SO_3 = C_{10}H_7O.SO_2NH_4 + NH_3 + H_2O,$$

the ester is then decomposed with reformation of ammonium sulphite:

$$C_{10}H_7O.SO_2NH_4 + 2NH_3 = C_{10}H_7NH_2 + (NH_4)_2SO_3.$$

If the ammonia is replaced by primary or secondary amines, secondary or tertiary amines are obtained.

Alkylnaphthylamines are obtained similarly to the alkylanilines.

The phenylnaphthylamines, α and β, are prepared from the corresponding naphthols by heating with aniline and aniline hydrochloride.

Diamines. The diamines that are important in the dyestuff industry are meta- and para-phenylene diamines and the corresponding tolulene diamines, benzidine, its homologues and derivatives, 1 . 5- and 1 . 8-naphthylene diamines, and para-amido-diphenylamine.

Meta-phenylene diamine (benzene ring with NH_2 at positions 1 and 3) (M.P. 83° C., B.P. 287° C.) is obtained by the reduction of m-dinitrobenzene under similar conditions to the preparation of aniline. It gradually turns brown when exposed to the air, and is soluble in acid, water, or alkaline carbonate solutions. It is employed as a developing agent and for the production of dyestuffs.

Para-phenylene diamine (benzene ring with NH_2 at positions 1 and 4) (M.P. 147° C., B.P. 267° C.) is obtained by the reduction of amino-azobenzol hydrochloride in dilute alcoholic solution or aqueous suspension by means of zinc dust, the temperature being maintained between 60° C. and 70° C. After reduction the solution is made alkaline with soda and the aniline removed by steam distillation. The liquor remaining is concentrated till the para-phenylene diamine crystallises out.

This method for the production of para-diamines is

of general application. P-phenylene diamine is used in the manufacture of sulphide dyes, and for the production of browns on fibres by means of oxidising agents.

Naphthalene diamines. Of the many possible isomers only the 1.5- and the 1.8-diamido naphthalenes are important. They are obtained from the corresponding dinitronaphthalenes by reduction with sodium or ammonium sulphide.

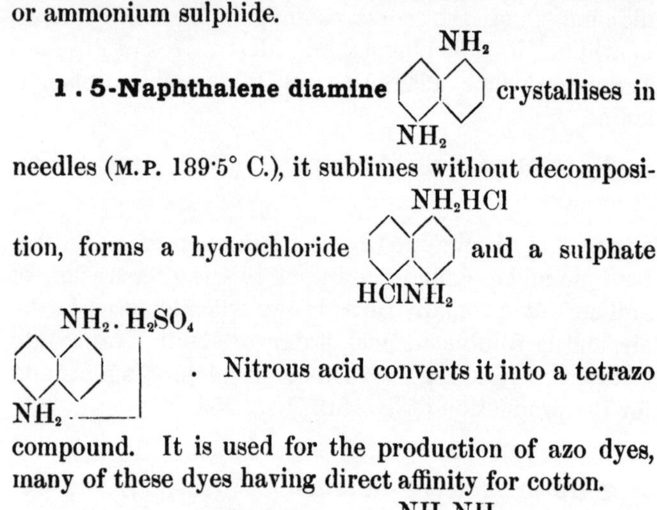

1.5-Naphthalene diamine crystallises in needles (M.P. 189·5° C.), it sublimes without decomposition, forms a hydrochloride and a sulphate. Nitrous acid converts it into a tetrazo compound. It is used for the production of azo dyes, many of these dyes having direct affinity for cotton.

1.8-Naphthalene diamine crystallises in needles (M.P. 67° C.) and forms salts similar to the 1.5-derivative. By treatment with nitrous acid a red azimide is formed

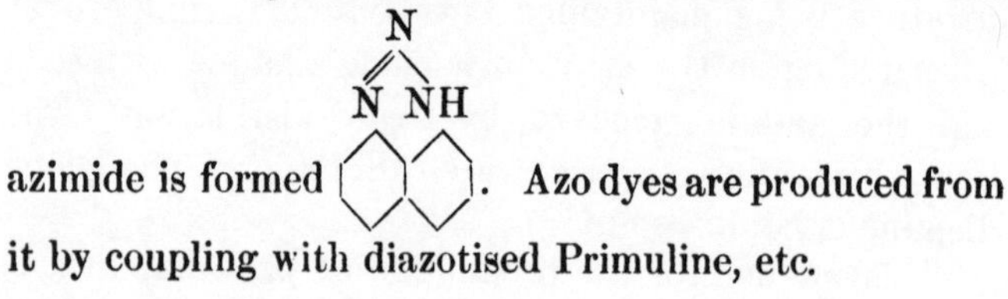

. Azo dyes are produced from it by coupling with diazotised Primuline, etc.

Many of the **naphthalene diamine sulphonic acids** are employed for the production of dyestuffs. These are obtained as follows:

(*a*) By reduction of azo derivatives of naphthylamine sulphonic acids.

(*b*) By heating the naphthol disulphonic acids with ammonia under pressure.

(*c*) Nitration and reduction of amino sulphonic acids.

(*d*) Action of sodium bisulphite upon naphthoquinone derivatives, etc.

Benzidine $H_2N\langle\ \rangle-\langle\ \rangle NH_2$ (M.P. 122° C.) is the mother substance of most direct cotton dyes and is obtained from nitrobenzene. Nitrobenzene is treated with zinc dust in alcoholic solution heated in a vessel provided with a reflux condenser, caustic soda is added to the solution, and the heating continued until the zinc has practically disappeared. The alcohol is distilled and the residue treated with hydrochloric acid, added in small quantities at a time, until an acid solution is obtained; the temperature must not exceed 35° C.; finally the solution is boiled, filtered hot, and the benzidine precipitated as sulphate by addition of sulphuric acid or sodium sulphate. Benzidine sulphate is only sparingly soluble in water. It is extensively used in the preparation of direct cotton dyes.

Tolidine $\begin{matrix}H_3C\\H_2N\end{matrix}\langle\ \rangle-\langle\ \rangle\begin{matrix}CH_3\\NH_2\end{matrix}$ (M.P. 129° C.) is obtained by a similar method from o-nitrotoluene, and **Dianisidine** $\begin{matrix}CH_3O\\H_2N\end{matrix}\langle\ \rangle-\langle\ \rangle\begin{matrix}OCH_3\\NH_2\end{matrix}$ (M.P. 168° to 172° C.) from o-nitroanisol. **Diphenetidine** is the corresponding compound obtained from o-nitrophenetol $C_6H_4 \langle {}^{OC_2H_5}_{NO_2}$.

By the nitration of diacetylbenzidine either mono- or dinitro-benzidine may be obtained. The diacetylbenzidine is dissolved in sulphuric acid with nitric acid added in quantity sufficient to form either the mono- or dinitro-compound. The product is precipitated by addition of water and the acetyl derivative hydrolysed by boiling with dilute hydrochloric acid, or strong caustic soda.

Dinitrobenzidine ${}^{O_2N}_{H_2N} \rangle C_6H_3-C_6H_3 \langle {}^{NO_2}_{NH_2}$ is a red compound (M.P. 218° to 221° C.).

Mono-nitrobenzidine $H_2N-C_6H_4-C_6H_3(NO_2)-NH_2$ (M.P. 143° C.) is a red compound obtained from benzidine, generally by the addition of the theoretical quantity of potassium nitrate to a sulphuric acid solution of benzidine. The nitro group being in the meta position to the amido group.

Dichlorobenzidine ${}^{Cl}_{H_2N} \rangle C_6H_3-C_6H_3 \langle {}^{Cl}_{NH_2}$ (M.P. 133° C.) is prepared by the direct chlorination of diacetylbenzidine and hydrolysis of the product with boiling hydrochloric acid.

Benzidinesulphone $H_2N-C_6H_3 \langle {}^{|}_{SO_2} \rangle C_6H_3-NH_2$ is obtained by heating benzidine sulphate with excess of fuming sulphuric acid (20 per cent. SO_3), the product is poured on to ice, filtered and separated from the accompanying benzidinesulphone mono- and di-sulphonic acids by means of caustic soda. Benzidine-sulphone melts above 350° C.

Benzidine dicarboxylic acid (Diamidodiphenic acid).

$$\begin{array}{ll} HOOC & COOH \\ H_2N & NH_2 \end{array}$$

Prepared by reduction of ortho-nitrobenzoic acid with zinc dust and caustic soda, followed by transformation of the hydrazobenzene derivative by boiling with dilute hydrochloric acid.

CHAPTER V

SULPHONIC ACIDS

THE production of sulphonic acids has been known since 1834. For the preparation of these compounds sulphuric, fuming sulphuric and chlorsulphonic acids are mainly used; in some cases sodium bisulphite is capable of producing sulphonic acids. The manner in which the sulphonation is carried out varies with the nature of the compounds to be treated. As with nitration, the sulphonation of phenols is generally more easily accomplished than the sulphonation of amines or hydrocarbons. The temperature at which sulphonation is done plays a very important part; not only does a rise of temperature aid sulphonation but it also influences the position of the sulphonic acid groups. Sulphonic acid groups often wander from their original position when the temperature is raised. This is also controlled to a certain extent by the groups already present in the compound. In some cases the introduction of sulphonic acid groups is to provide intermediate steps in the production of other compounds, e.g., nitro bodies, but they are generally introduced for the purpose of producing water soluble bodies or giving acid properties. The sulphonic acids are often difficult to obtain pure and therefore to identify. The commercial products are usually amorphous bodies obtained by evaporation of solutions to dryness (Fig. X, Appendix). The most characteristic reactions of amido and oxy-sulphonic acids are obtained with diazo-compounds (see Azo Dyes). The sulphonic acids rarely melt sharply on heating or crystallise well from solutions. Titration with alkalis, boiling with dilute acid to remove sulphonic acid groups, or treatment with phosphorus pentachloride to obtain chlorine compounds of definite melting point are the chief methods available. The commercial sulphonic acids are generally sold as sodium salts.

Benzene monosulphonic acid ⬡ SO_3H is of little importance, but on further sulphonation **benzene disulphonic acid** is obtained from which resorcin is manufactured. In time of war the price of phenol may be so high that it pays to manufacture it from benzene by sulphonation followed by decomposition of the sulphonate with caustic alkali.

Benzene disulphonic acid is obtained by heating benzene with fuming sulphuric acid at about 100° C. in iron vessels provided with stirring gear until the benzene has disappeared; the temperature is then gradually raised to 275° C. and the heating continued for three hours (Fig. IV, Appendix). The solution, after cooling, is diluted and treated with chalk, filtered, and the filtrate treated with potassium or sodium carbonate to prepare the potassium or sodium salt of the disulphonic acid. The three isomeric disulphonic acids of benzene are produced, but mainly the meta derivative. All on heating with caustic soda give resorcin.

The crude salt may be fused with caustic soda for the preparation of resorcin, or the solution of the potassium salt concentrated to 1·275 sp. gr. and allowed to crystallise. The potassium-meta-disulphonate separates out, leaving the potassium-para-disulphonate in solution.

Piria some forty years ago discovered that α-nitronaphthalene, on treating with aqueous alcoholic ammonium sulphite, gave thionaphthamic acid and naphthionic acid. With a large excess of bisulphite, α-naphthylamine 2.4-disulphonic acid is produced instead of **Piria's compounds.** The use of bisulphites in sulphonation is restricted by their simultaneous reducing action. They are used for the production of naphthylamine-3.8-disulphonic acid, and for reduction and sulphonation of 1.5- and of 1.8-dinitronaphthalene. In these cases nitrogen is not eliminated from the molecule, but with 1.8-dinitro-3.6-disulphonic acid one of the nitro groups is replaced by the hydroxy group.

Naphthalene sulphonic acids may be obtained

(*a*) by sulphonation of naphthalene;

(*b*) by the elimination of the amido group from the amido sulphonic acids by diazotising etc.;

(*c*) by replacement of the amido group for the thio-oxy group, and oxidation of the latter group into a sulphinic and finally sulphonic acid.

Of these methods the sulphonation of naphthalene

is the most important. Unlike most substituting agents sulphuric acid gives rise to the production of β-derivatives in large amount. In no case however have two sulphonic acid groups been introduced in positions relatively 1.2, 1.4, or 1.8 to one another. The sulphonic acids and their alkali salts are readily soluble in water, but the lead, barium, and calcium salts differ in solubility and often afford a means of separating the different compounds from a sulphonation product.

From the alkali salts, crystalline amides and anilides may be prepared which, having definite melting points, serve to characterise the acids; from the sulphonic chlorides, chloronaphthalenes may be produced which help to determine the constitution of the original acids.

By hydrolysis with caustic alkalis the naphthalene sulphonic acids give rise to hydroxy compounds, in which one or more sulphonic acid groups may be replaced by oxy groups.

The sulphonic acid groups are more easily eliminated from the α- than from the β-positions, and in some cases further change may take place, e.g., a sulphonic acid group in the meta position to a hydroxy group may give a cresol by caustic hydrolysis, e.g.[1],

S S S S → S OH S S

↘ HO OH OH → HO CH_3

The position in which the sulphonic acid group enters

[1] The symbol S is here used in place of SO_3H and this well-known abbreviation is employed frequently where it is quite obvious what it stands for.

the naphthalene ring depends to a large extent upon the temperature at which the reaction takes place. The sulphonic acid group may also wander from the α- to the β-position by the action of heat.

The following diagrammatic scheme shows how the different sulphonic acids are obtained from one another, S representing the sulphonic acid group.

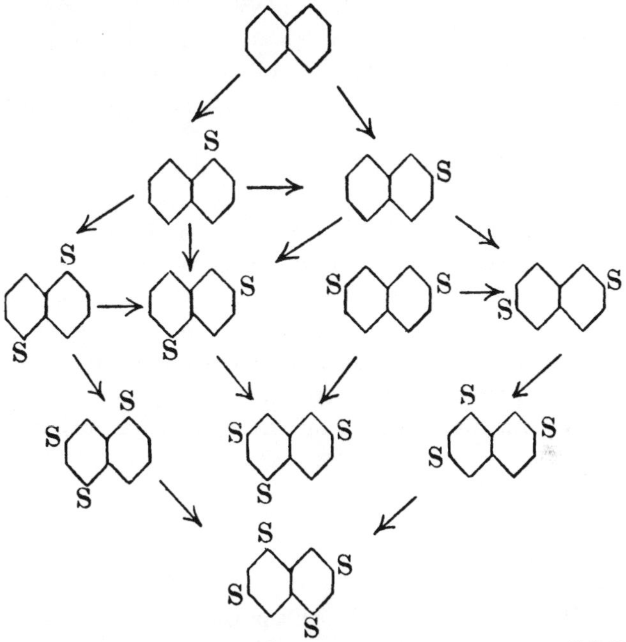

α-naphthalene sulphonic acid (SO_3H in the α-position).

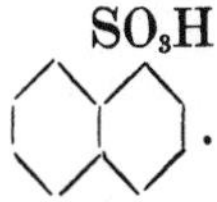

Finely powdered naphthalene is heated with half its weight of concentrated sulphuric acid at 40° C. until the product is completely soluble in caustic soda solution.

The reaction mass on cooling separates into a lower layer of sulphuric acid and an upper layer of α-naphthalene sulphonic acid, which crystallises out with one molecule of water. M.P. 85° to 90° C.

The acid may be purified or directly converted into α-naphthol by alkali fusion.

β-naphthalene sulphonic acid 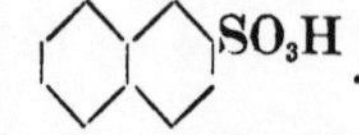

.

Equal weights of concentrated sulphuric acid and naphthalene are heated to 180° C. for four to five hours. The product, after cooling, is run into an excess of brine, the sodium salt of naphthalene β-sulphonic acid separates out, it is filter pressed, dried, and ground. This compound is not so readily obtained pure as the α-acid. It is used for the manufacture of β-naphthol.

The introduction of a second sulphonic acid group into naphthalene also takes place at low temperatures in the α- and at higher temperatures in the β-positions, the second group entering the molecule as far away as possible from the group already present.

Naphthalene-1.5-disulphonic acid (Armstrong's acid) S S is obtained by sulphonation of naphthalene at 70° C. with fuming sulphuric acid.

Naphthalene-1.6-disulphonic acid is obtained by sulphonation of the naphthalene-β-sulphonic acid with fuming sulphuric acid at 100° C.

Naphthalene-1.6-, -2.6-, and -2.7-disulphonic acids are obtained by sulphonation of naphthalene with excess of sulphuric acid at 180° C. The calcium salts are formed by addition of lime, and these can be separated by means of brine. The calcium salt of the 1.6 acid is fairly soluble in cold brine, the 2.7 salt is tolerably soluble in hot but sparingly soluble in cold, and the 2.6 salt is practically insoluble in hot and cold brine.

Naphthalene trisulphonic acids are produced by the further sulphonation of naphthalene disulphonic acids or their salts by means of fuming sulphuric acid or chlorosulphonic acid. The most important are the 1.3.5, the 1.3.6, the 1.3.7, and the 2.3.6 trisulphonic acids. When fused with alkali they give oxynaphthalene sulphonic acids. Tetrasulphonic acids are also produced by sulphonation of naphthalene disulphonic acids.

Amido and oxy-naphthalene sulphonic acids.

The naphthylamine sulphonic acids may be obtained by the following methods:

(*a*) By the sulphonation of naphthylamine.

(*b*) By the nitration of naphthalene sulphonic acids, followed by reduction.

(*c*) By the further sulphonation of the acids obtained by method (*b*).

(*d*) By the replacement of α-oxy groups of naphthol sulphonic acids by the amido group, e.g., heating with ammonia in the presence of ammonium sulphite.

As in the case of the naphthalene sulphonic acids, the sulphonic acid group may wander in the molecule. By heating naphthionic acid (sodium salt) it is converted into the sodium salt of α-naphthylamine-2-sulphonic acid.

The sulphonic acid groups in the α-positions are more easily eliminated than those in the β-positions, whether the agent be water, caustic alkali, or dilute acid at high temperature.

Naphthylamine sulphonic acids.

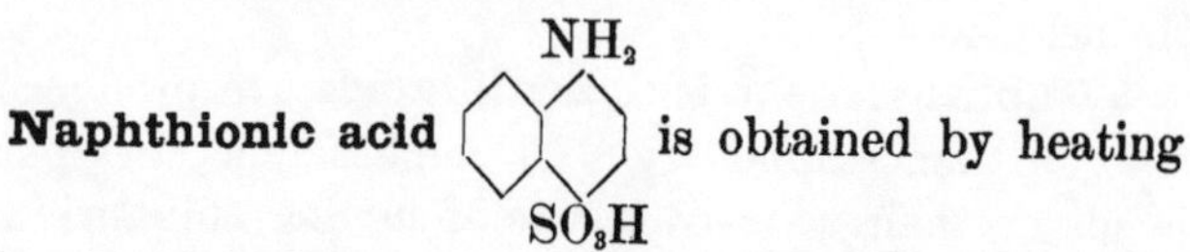

Naphthionic acid is obtained by heating molecular quantities of concentrated sulphuric acid and α-naphthylamine at 60° C., the additive sulphate formed being then mixed with about 3 per cent. of oxalic acid and baked at 180° C. during eight hours. The oxalic acid decomposes on heating into carbon dioxide and water which causes the mass to swell and become porous, thus aiding the reaction. The free acid is sparingly soluble in water, while the commercial product, which is the sodium salt, is more soluble, cf. sulphanilic acid, p. 32.

1 . 5-Naphthylamine sulphonic acid (Laurent's acid) (NH_2, SO_3H) is obtained by sulphonating α-naphthylamine with fuming sulphuric acid at 0° C. or by reducing α-nitronaphthalene-5-sulphonic acid. The acid separates as a slime on pouring the product of the reaction on ice.

2 . 5-Naphthylamine sulphonic acid (Dahl's acid) (NH_2, SO_3H) is obtained by mixing β-naphthylamine sulphate with cold concentrated sulphuric acid and allowing to stand at ordinary temperature for two or three days.

2 . 6-Naphthylamine sulphonic acid (Brönner's acid) H_2N-naphthalene-SO_3H is obtained as the ammonium salt by heating 2 . 6-naphthol sulphonic acid with excess of ammonia in an autoclave at 190° to 200° C. for two days.

2 . 7-Naphthylamine sulphonic acid (F-acid) H_2N-naphthalene-SO_3H is obtained from 2 . 7-naphthol sulphonic acid by heating with ammonia and ammonium sulphite in an autoclave. It is obtained along with Brönner's acid by heating β-naphthylamine and concentrated sulphuric acid at 170° C.

Naphthylamine disulphonic acids are obtained by the further sulphonation of the monosulphonic acids by means of fuming sulphuric acid, or by the replacement of the oxy group in the corresponding naphthol disulphonic acids by the amido group. This is generally carried out by heating with ammonia under pressure in presence of ammonium sulphite.

α-naphthylamine-4.6- and **-4.7-disulphonic acids** are obtained by sulphonating naphthionic acid with 25 per cent. (SO_3) fuming sulphuric acid at about 30° C. The mixed acids are converted into calcium salts, dried, and extracted with 85 per cent. alcohol, the calcium salt of the 4 . 6-acid being soluble.

β-naphthylamine-3 . 6- (**amido-R-acid**) and **-6 . 8-disulphonic acids** (**amido-G-acid**) are obtained by the amidation of corresponding naphthol disulphonic acids with ammonia in presence of ammonium sulphite.

The naphthol sulphonic acids are obtained by the following general methods :

(1) Sulphonation of naphthol.

(2) Substitution of the amido group for the hydroxyl group by the action of bisulphite or by the action of water at a high temperature.

(3) Replacement of chlorine in α-chloronaphthalene sulphonic acids by action of caustic alkali.

(4) Replacement of a sulphonic acid in naphthalene polysulphonic acids by action of caustic alkali.

The direct sulphonation of the naphthols is of most importance for the derivatives of β-naphthol, the α-naphthol sulphonic acids being seldom prepared by this method. This is no doubt due to the ease with which sulphonation takes place in the α-position compared with the β-substitution.

The temperature at which sulphonation is carried out, and the duration of heating, are important factors in the production of these bodies on account of the alteration in the position of the sulphonic acid group brought about by heating many of these acids.

Naphthol sulphonic acids. The mono-, di- and tri-sulphonic acids of alpha- and beta-naphthol are very important intermediate bodies in the production of azo dyestuffs. A very interesting reaction which is often employed in their preparation involves the interchange of the amido group and the oxy group. In presence of ammonium sulphite the oxy group is replaced by the amido group, while in the presence of sodium bisulphite the amido group is replaced by the oxy group.

α-naphthol-4-sulphonic acid (**"N.W."** or **Neville and Winther's acid**). Sodium naphthionate is heated with eight times its weight of sodium bisulphite solution of 76° Tw., and twice its weight of water, in a closed vessel, until the solution no longer gives a precipitate of naphthionic acid on addition of hydrochloric acid.

α-naphthol-5-sulphonic acid (**Clevé's acid**) is obtained similarly from the corresponding amido acid.

β-naphthol-8-sulphonic acid (**Crocein** or **Bayer's acid**). β-naphthol in a finely powdered condition is mixed with twice its weight of concentrated sulphuric acid, the temperature not being allowed to exceed 20° C. The mixture is tested after about seven days' standing by boiling with an equal volume of water to find when the naphthol is no longer precipitated.

β-naphthol-6-sulphonic acid (**Schäffer's acid**) is obtained by sulphonating as above but at 50° to 60° C. A mixture of about equal parts of Bayer's and Schäffer's acid is obtained. About 80 per cent. of the Schäffer's acid is obtained as sodium salt by adding about 66 per cent. of the theoretical quantity of caustic soda and allowing to crystallise.

β-naphthol-6 . 8-disulphonic acid, "G-acid"

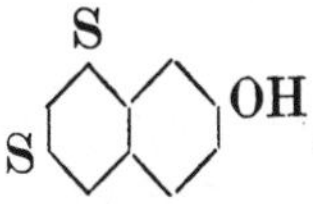

S

S OH, and **β-naphthol-3 . 6-disulphonic acid,**

"R-acid" S OH S, are obtained by sulphonating β-naphthol with four times its weight of sulphuric acid, the G-acid mainly at 60° C. and the R-acid mainly at 100° C. They are separated from one another as sodium salts by means of 20 per cent. salt solution, the R-salt being the more sparingly soluble.

Amidonaphthol sulphonic acids.

2 . 8-Amidonaphthol-6-sulphonic acid is obtained by heating amido-G-acid with twice its weight of 50 per cent. caustic soda solution for six hours at 185° C. (Fig. VI, Appendix).

1.8 Amidonaphthol-3.6-disulphonic acid, "H-acid" (OH NH_2 / S, S), is obtained by heating the corresponding diamido acid with 40 per cent. caustic soda to 200° C. in an autoclave. The acid and the acid salts are sparingly soluble in water.

Further information may be obtained from an excellent article in Thorpe's *Dictionary of Applied Chemistry* under "Naphthalene."

Diamido-disulphonic acids.

Diamidostilbene-disulphonic acid

$$H_2N\langle S \rangle CH = CH\langle S \rangle NH_2$$

is obtained by boiling the sodium salt of p-nitrotoluene sulphonic acid with 33 per cent. caustic soda, in this manner dinitrostilbene-disulphonic acid is obtained. The diamido acid is obtained by reducing the alkaline solution at the boil with zinc dust until the solution no longer turns red on exposure to air, it is then filtered into an excess of hydrochloric acid.

CHAPTER VI

HALOGEN COMPOUNDS

THE chloro-derivatives of the aromatic compounds are in most cases easily obtained by the action of chlorine in presence of a catalyst. The most important catalytic agents are ferric chloride, aluminium chloride, molybdic chloride ($MoCl_5$) and iodine.

In the cold and dark the chlorination takes place mainly in the ring, at the boiling point and in sunlight the chlorination of homologues, e.g., toluene, occurs mainly in the side chain. These compounds are produced mostly for the manufacture of aldehydes and carboxylic acids.

The manufacture of caustic soda by electrolysis of salt has given a cheap supply of chlorine. This has increased the use of chloro-derivatives such as chloracetic acid, phosgene, chlorobenzaldehydes etc., and has enabled the use of chlorobenzidine etc., to be extended giving dyes of greater fastness.

Chloronaphthalene sulphonic acids may be employed for the preparation of naphthylamine sulphonic acids by means of ammonia in presence of copper salts. Chlor- and brom-anthracene derivatives are also of importance for the production of alkyl and aryl derivatives of anthracene etc.

Chlorobenzene (benzene ring with Cl) (B.P. 129° C.) is obtained by action of chlorine upon benzene in presence of ferric chloride. The product is then fractionated.

By further chlorination **para-dichlorobenzene** (structural formula: benzene ring with Cl at top and bottom) (M.P. 53° C.) is obtained, this compound is employed for the production of p-phenylene diamine by heating under pressure with ammonia in the presence of a copper salt.

Ortho-chlorotoluene (B.P. 159° C.) and **para-chlorotoluene** (B.P. 162° C., M.P. 7·4° C.) are obtained by chlorination of toluene in the cold in the presence of molybdic chloride ($MoCl_5$).

Benzyl chloride $C_6H_5CH_2Cl$ (B.P. 176°—179° C.) is obtained by chlorination of toluene at the boil, in earthenware vessels, heated with a leaden steam pipe. The vapours are condensed and returned to the vessel. The chlorination is continued until a sample has the specific gravity of 1·1 at 15° C. The product is then fractionated, some unchanged toluene and benzal chloride distil over.

Benzal chloride $C_6H_5CHCl_2$ (B.P. 206° C.) is obtained by further chlorination until the specific gravity is about 1·26. The product is then rectified.

Benzotrichloride $C_6H_5CCl_3$ (B.P. 213°—214° C., Sp. Gr. 1·38) is obtained as a by-product in the production of benzal chloride.

α-Chloronaphthalene $C_{10}H_7Cl$ (B.P. 252° C.) is obtained by chlorination of naphthalene at the boil in presence of ferric chloride. The product is then purified by distillation.

The **dichloronaphthalenes** 1.4, and 1.5, are obtained by chlorination in carbon disulphide solution below 0° C., traces of the 1.2, and 1.7 isomers are also present.

Tetrachloronaphthalene (structural formula: naphthalene ring with four Cl) is obtained by action of chlorine upon naphthalene in shallow layers in the cold, the product is extracted with petrol to remove the dichloronaphthalenes.

A large number of chloronaphthalene sulphonic acids is obtained by sulphonation of the chloronaphthalene derivatives.

Chlorophthalic anhydrides are obtained by the chlorination of phthalic anhydride in fuming sulphuric acid solution.

$$C_6H_4{<}(CO)_2{>}O + 2Cl_2 + 2SO_3 = C_6H_2Cl_2{<}(CO)_2{>}O + 2SO_3HCl.$$

Three isomerides are obtained, the main compound being the 3.6-dichlorophthalic anhydride. The three isomerides may be separated by crystallisation of the zinc salt of the dichlorophthalic acid.

Tetrachlorophthalic acid and **anhydride**

$$C_6Cl_4(COOH)_2 \text{ and } C_6Cl_4{<}(CO)_2{>}O$$

are obtained by passing chlorine through a mixture of antimony chloride and phthalic anhydride heated to 200° C. When chlorine is no longer absorbed, the antimony pentachloride is distilled off, the residue may be either distilled, or well washed with water after cooling. The anhydride is insoluble in cold water (M.P. 252° C.).

CHAPTER VII

PHENOLIC COMPOUNDS

ALTHOUGH phenolic substances are present in coal tar, the majority of the hydroxy compounds other than phenol and cresols are prepared from the corresponding sulphonic acids or chloro derivatives by fusion with alkali.

$$R . SO_3Na + 2NaOH = R . ONa + Na_2SO_3.$$

Phenol is usually extracted from coal tar but may be manufactured from benzene. Hydroxy compounds are sometimes obtained from corresponding amino compounds, either by diazotisation and subsequent decomposition of the diazo compound by water, or by the reaction of sodium bisulphite upon the amine followed by treatment with caustic soda. This method is more common with the naphthalene derivatives. Oxysulphonic acids may, in some cases, be prepared by heating amines under pressure with sodium bisulphite solution.

Resorcin (benzene ring with OH and OH) or **Resorcinol** (M.P. 118° C., B.P. 276·5° C.) is manufactured by heating sodium benzene disulphonate with caustic soda in cast iron vessels, fitted with a stirring gear, at 270° C., for about nine hours. The melt is dissolved in water and acidified with sulphuric acid. The resorcin is extracted with ether (Fig. IX, Appendix) and the ether distilled off. The crude resorcin may be purified by recrystallisation, quick distillation under reduced pressure, or sublimation (Fig. VII, Appendix).

Pyrogallol $C_6H_3(OH)_3$ [benzene ring with three adjacent OH groups] (M.P. 131° C.). This compound is not obtained from coal tar. It is produced from tannin by hydrolysing to **gallic acid** $C_6H_2(OH)_3COOH$ [benzene ring with OH, HO, OH and COOH], the gallic acid is then heated with twice its weight of water in an autoclave from 200° to 210° C. The solution obtained of the very soluble pyrogallol is decolourised by animal charcoal and evaporated.

Since carbon dioxide is so readily split off from gallic acid, it is often better in high temperature reactions to use gallic acid in place of pyrogallol, the latter body being produced during the condensation by loss of carbon dioxide.

Naphthols $C_{10}H_7OH$.

α-**naphthol** (M.P. 94° C., B.P. 280° C.),

β-**naphthol** (M.P. 122° C., B.P. 286° C.).

These are white crystalline compounds, sparingly soluble in water and unlike phenol do not absorb water. They are manufactured from the corresponding sulphonic acids. Caustic soda is dissolved in water until the solution has a boiling point of about 130° C., into this solution finely powdered dried sodium naphthol sulphonate is stirred and the mixture gradually heated to about 300° C.; the reaction being carried out in an autoclave. The melt, after cooling, is acidified with hydrochloric acid and the naphthol which separates out is filter-pressed, washed and then melted.

The dioxynaphthalenes are obtained by an alkali melt of the corresponding disulphonic acids or oxy-sulphonic acids.

The replacement of sulphonic acid groups by oxy groups is easier in the α position and takes place at a lower temperature than in the β position. On account of this difference it is possible to obtain oxysulphonic acids by fusion processes from disulphonic acids.

Chloronaphthols may be obtained from the naphthols by chlorination in cooled acetic acid solution, by this method only homonucleal derivatives can be obtained. Other chloro-derivatives are obtained by the Gattermann or Sandmeyer reactions from the corresponding amino derivatives. Some of the chloronaphthol sulphonic acids are employed in the production of azo dyes, the introduction of chlorine tending to improve general fastness.

The **Nitronaphthols** may be obtained by the direct nitration of the naphthols, from nitrosonaphthols by oxidation, from naphthol sulphonic acids, or from nitrosonaphtholsulphonic acids. In many cases the yields are very unsatisfactory owing to the formation of tar. Chloronitronaphthalenes are converted into naphthols by digesting with strong soda carbonate solution, just as benzene derivatives are converted into phenols.

Dioxynaphthalene derivatives. Like the naphthols, the dioxynaphthalenes may be obtained by heating sulphonic acids with caustic alkali, but it has not been found possible to obtain 1 . 2-, 1 . 3-, or 1 . 4-dioxynaphthalene by this method.

The dioxynaphthalenes may also be obtained by

heating diamidonaphthalenes, amidonaphthols, certain amidonaphthol sulphonic acids, or certain dioxynaphthalene sulphonic acids under pressure with dilute mineral acids. In the case of the sulphonic acid derivatives, the sulphonic acid group is eliminated by hydrolysis.

The 1.2- and the 1.4-dioxynaphthalenes are only obtained by reduction of the corresponding quinones.

The alkaline solutions of the dioxynaphthalenes absorb oxygen from the air forming dark brown or black solutions.

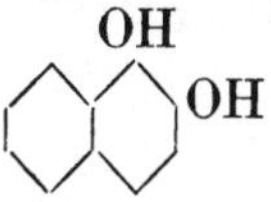

OH

OH **1.2-Dioxynaphthalene** (M.P. 60° C.)

is obtained by reducing β-naphthoquinone with a cold solution of sulphurous acid. It readily oxidises back to β-naphthoquinone by treatment with ferric chloride solution.

It forms scales, dissolves in alkalis forming yellow solutions which turn green on exposure to air.

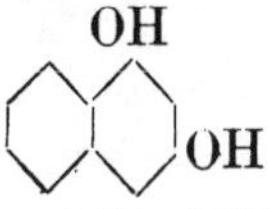

OH

OH **1.3-Dioxynaphthalene** (M.P. 124° C.; diacetate, M.P. 56° C.) is obtained from 1-amido-3-naphthol by boiling with dilute acid, or from many of the amido-oxynaphthol sulphonic acids, having the amido and oxy groups in the 1.3 positions to one another, by heating under pressure with water or dilute acid.

It crystallises from water in scales.

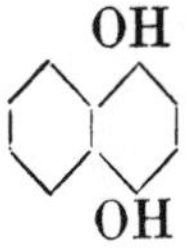

OH

1.4-Dioxynaphthalene (α-naphtho-hydro-

OH

quinone) (M.P. 176° C.; diacetate, M.P. 128°—130° C.) is

obtained by reduction of α-naphthoquinone with tin and boiling hydrochloric acid.

It is readily soluble in boiling water.

OH

OH

1 . 5-Dioxynaphthalene (M.P. 265° C.; diacetate, M.P. 159°—160° C.) is obtained by caustic fusion of the corresponding naphthol sulphonic acid.

It readily oxidises in alkaline solution.

1 . 5-Dioxynaphthalene-2-sulphonic acid ("C-acid") is obtained together with the -4-sulphonic acid when 1.5-dioxynaphthalene is heated with sulphuric acid at 50°—60° C.

It couples in alkaline solution with diazo salts to form azo dyestuffs.

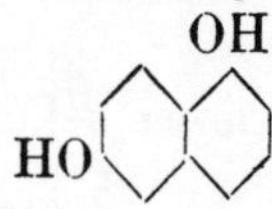

1 . 6-Dioxynaphthalene (M.P. 135° C.; diacetate, M.P. 73° C.) is obtained by caustic fusion at 230°—250° C. of the corresponding naphthalene disulphonic acid, or when β-naphthol-5-sulphonic acid is similarly treated.

It crystallises from benzene.

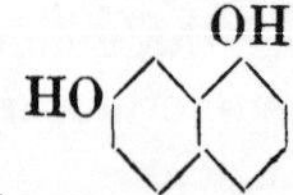

1 . 7-Dioxynaphthalene (M.P. 170° C.; diacetate, M.P. 108° C.) is obtained from β-naphthol-8-sulphonic acid by caustic fusion.

1 . 7-Dioxynaphthalene-3-sulphonic acid is obtained by heating β-naphthol disulphonic acid G with caustic soda at 220°—230° C.

It is used for the production of azo dyestuffs.

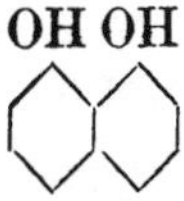

OH OH

1.8-Dioxynaphthalene (M.P. 140° C.; diacetate, M.P. 147° C.) is obtained from α-naphthol-8-sulphonic acid by caustic fusion, or from naphthosultone by caustic fusion at 220°—230° C. **Naphthosultone**

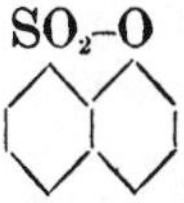

SO_2–O

is obtained almost quantitatively by boiling diazotised α-naphthylamine-8-sulphonic acid with water, alcohol, or dilute sulphuric acid.

The 1.8-dioxynaphthalene. It is also obtained when 1.8-diamido-naphthalene is heated with dilute hydrochloric acid at 180° C.

It oxidises rapidly when moist on exposure to air.

1-8-Dioxynaphthalene-3-sulphonic acid is obtained by heating α-naphthol-6.8-disulphonic acid with 50 per cent. caustic soda solution at 170°—210° C.

It couples in alkaline solution with diazo salts.

1.8-Dioxynaphthalene-3.6-disulphonic acid

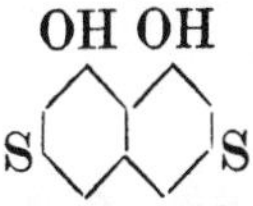

OH OH

S S (**"Chromotropic acid"** or **"Chromogen I"** (M.)) is obtained from Koch's α-naphthol-3.6.8-trisulphonic acid by heating with 60 per cent. caustic soda at 170°—220° C.; or from 1-amido-8-naphthol-3.6-disulphonic acid with 5 per cent. caustic soda solution at 265° C.; also from 8-chloro-1-naphthol-3.6-disulphonic acid by heating with 60 per cent. caustic soda solution at 220°—240° C.

Chromogen I is used as a dyestuff on wool, being absorbed from an acid bath, and on subsequent oxidation with bichromate giving a fast brown. Chromotropic acid is also largely used in the production of mono- and disazo dyestuffs, having mordant dyeing properties.

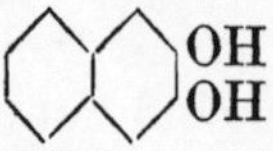

2 . 3-Dioxynaphthalene (M.P. 160°–161° C.) is obtained by heating 2-amido-3-naphthol-6-sulphonic acid with dilute mineral acid under pressure at 180° to 200° C., or by heating 2 . 3-dioxynaphthalene-6-sulphonic acid with 25 per cent. sulphuric acid to 200° C., or by fusion of β-naphthol-3 . 6-disulphonic acid with strong caustic soda at 280° to 300° C.

It is sparingly soluble in water.

2 . 3 Dioxynaphthalene-6 . 8-disulphonic acid is obtained by fusion of β-naphthol-3 . 6 . 8-trisulphonic acid with caustic potash.

It readily condenses to form azo compounds.

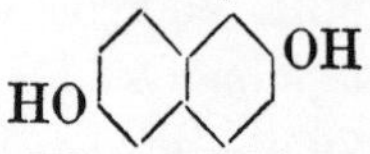

2 . 6 - Dioxynaphthalene (M.P. 218° C.; diacetate, M.P. 175° C.) is obtained from the corresponding naphthalene disulphonic acid or the corresponding naphthol monosulphonic acid by fusion with caustic soda.

It is sparingly soluble in water.

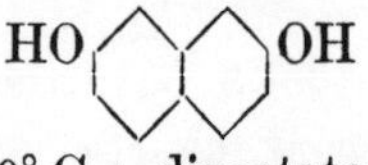

2 . 7 - Dioxynaphthalene (M.P. 190° C.; diacetate, M.P. 136° C.) is formed similarly to the 2 . 6-compound from 2 . 7-substituted derivatives.

It is readily soluble in hot water.

2 . 7-Dioxynaphthalene-3 . 6-disulphonic acid is obtained by sulphonation of the preceding compound with sulphuric acid at 100° C.

It couples in alkaline solution with one molecule of a diazo salt.

Amidophenols.

Ortho-amidophenol NH_2 OH (M.P. 170° C.) is prepared by the reduction of ortho-nitrophenol by means of sulphuretted hydrogen in presence of ammonia. It is a crystalline solid.

Many derivatives of ortho-nitrophenol are employed as first components in the preparation of azo and mordant azo dyestuffs.

Picramic acid. See page 26.

Para-nitro-ortho-amidophenol OH NH_2 NO_2 is obtained by the partial reduction of 2.4-dinitrophenol by means of ammonia and sulphuretted hydrogen.

Ortho-nitro-ortho-amidophenol-para-sulphonic acid OH O_2N NH_2 SO_3H is obtained by the nitration and partial reduction of phenol-para-sulphonic acid.

Ortho-chloro-ortho-amidophenol-para-sulphonic acid OH Cl NH_2 SO_3H is made from the preceding substance by diazotisation, halogen substitution by the Gattermann reaction, and reduction of the product.

Para-chloro-ortho-amidophenol-ortho-sulphonic acid is prepared by the action of sodium sulphite upon para-chloro-ortho-nitrophenol.

Meta-amidophenol $C_6H_4(OH)(NH_2)$ (M.P. 121° C.) may be prepared by heating 10 parts of resorcin under pressure with 20 parts of 10 per cent. aqueous ammonia, in presence of 6 parts of ammonium chloride, for twelve hours in an autoclave at 200° C. It may also be prepared by heating meta-sulphanilic acid with twice its weight of caustic soda at 280° to 290° C.

It is a colourless crystalline body, easily soluble in hot water, moderately in cold.

Dimethyl-meta-amidophenol $C_6H_4(OH)N(CH_3)_2$ (M.P. 87° C., B.P. 266° C.) may be prepared by heating dimethyl-meta-sulphanilic acid with caustic alkali as above. It may also be prepared from resorcin and dimethylamine.

It is a colourless crystalline body nearly insoluble in water.

Monoethyl-meta-amidophenol $C_6H_4(OH)NHC_2H_5$ (M.P. 62° C.) is obtained as above from monoethyl-meta-sulphanilic acid. The monoethyl-meta-sulphanilic acid is obtained from monoethylaniline by sulphonation with fuming sulphuric acid (30 per cent. SO_3).

It is a colourless crystalline body.

Diethyl-meta-amidophenol $C_6H_4(OH)N(C_2H_5)_2$ (M.P. 14° C.) is obtained similarly to the dimethyl compound.

It is a crystalline solid.

Phenyl-meta-amidophenol 

(M.P. 82° C.) or meta-oxy-diphenylamine is obtained by heating 12 parts of aniline hydrochloride with 10 parts of meta-amidophenol in an autoclave at 210° to 215° C.

It is a colourless crystalline solid, moderately soluble in boiling water.

Dimethyl-meta-amido-para-cresol 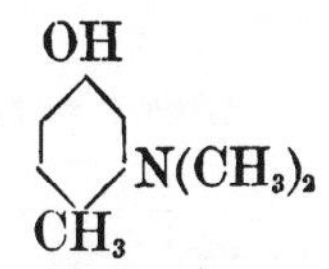

(M.P. 46° C.) is obtained by sulphonation of dimethyl-orthotoluidine with fuming sulphuric acid in the cold and decomposition of the sulphonic acid by heating with caustic soda as above. It is a crystalline powder.

Diethyl-meta-amido-para-cresol 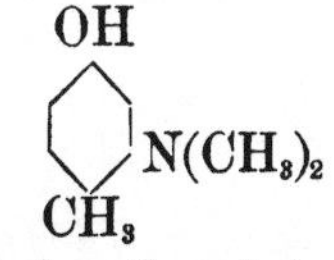

(M.P. 49° C.) is obtained as above from diethyl-orthotoluidine. It is a crystalline solid.

The meta-amido-phenols are used in the preparation of basic dyes.

OH

Para-amidophenol (M.P. 184° C.) is prepared

NH_2

by the electrolytic reduction of nitrobenzene in sulphuric acid solution or suspension. The compound formed is resolved on heating into para-amidophenol. It is also prepared by heating para-chloronitrobenzene under pressure with ammonia.

It crystallises in plates, and is readily oxidised. It is largely used in the preparation of sulphide blacks.

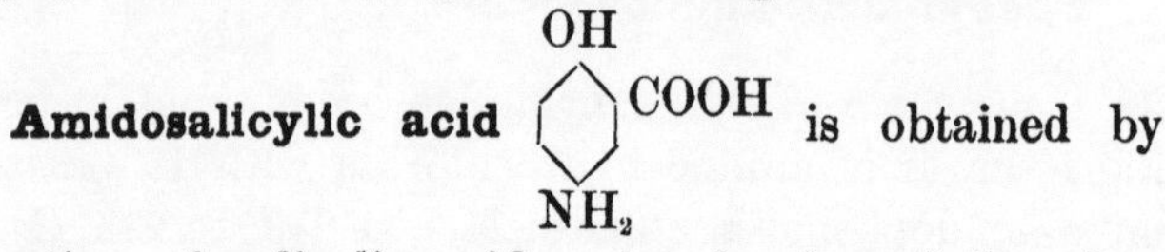

Amidosalicylic acid is obtained by nitration of salicylic acid and reduction of the nitro compound. It is sparingly soluble in water.

Meta-nitro-para-amidophenol (OH; NO_2; NH_2) (M.P. 148° C.) with decomposition, is obtained by hydrolysis of the acetyl derivative with dilute sulphuric acid. The acetyl para-amidophenol is nitrated with strong nitric acid.

Ortho-nitro-para-amidophenol-ortho-sulphonic acid (OH; O_2N; SO_3H; NH_2) is obtained by the nitration of para-amidophenol sulphonic acid. It is almost insoluble in cold water but forms an easily soluble diazo salt.

Amidonaphthols and their sulphonic acids.

1-Amido-2-naphthol (NH_2; OH) is produced by the reduction of nitroso-β-naphthol with ammonium sulphide. It is sparingly soluble in water, and is readily oxidised in alkaline solution. Ferric chloride and acid oxidising agents convert it into β-naphthoquinone.

It has been used for the production of oxazine

dyestuffs, as a coupling agent for azo dyestuffs and also for the production of azo dyestuffs by diazotisation in neutral solution in presence of copper salts.

1-Amido-2-naphthol-4-sulphonic acid (structural formula: naphthalene ring with NH_2, OH, S) is obtained by reduction of nitroso-β-naphthol with sulphurous acid. It may be diazotised in neutral solution in presence of copper salts and is used for the production of azo dyestuffs.

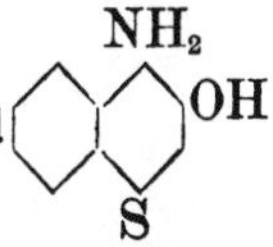

1-Amido-4-naphthol (structural formula: naphthalene ring with OH, NH_2) may be prepared by reduction of nitroso-α-naphthol. Its N-acetyl derivative, obtained by acetylating the hydrochloride in presence of sodium acetate is used as a coupling agent in alkaline solution.

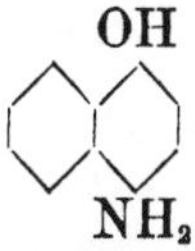

Several sulphonic acids of the above compound have been obtained but find little or no application in the production of dyestuffs.

1-Amido-5-naphthol (structural formula: naphthalene ring with NH_2, OH) may be prepared from α-naphthylamine-5-sulphonic acid by heating with a 60 per cent. solution of caustic soda at 240° to 250° C.

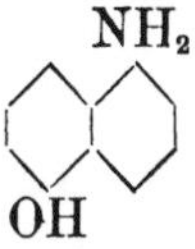

1-Amido-5-naphthol-2-sulphonic acid may be obtained by heating 1.5-diamido-naphthalene-2-sulphonic acid with sodium bisulphite solution followed by decomposition of the resulting body by treatment with caustic soda. It forms sparingly soluble diazo compounds.

1-Amido-5-naphthol-7-sulphonic acid (NH_2, S, OH)

(**M-acid**) is obtained by heating 1.5-dioxy-naphthalene-7-sulphonic acid with ammonium sulphite and ammonia under pressure. It is used as a coupling agent and also for diazotisation.

1-Amido-6-naphthol (NH_2, HO) may be obtained by reduction of 5-nitro-β-naphthol, or by heating β-naphthol with sodamide and naphthalene to 230° C. It may be diazotised and it also couples in alkaline solution with diazo salts.

1-Amido-6-naphthol-3-sulphonic acid and **1-amido-6-naphthol-4-sulphonic acid** may be used for the production of azo dyestuffs.

1-Amido-7-naphthol (NH_2, HO) may be obtained by digesting α-naphthylamine-7-sulphonic acid with 60 per cent. caustic soda solution at 250° C. It may be diazotised and it couples in acid or alkaline solutions with diazo salts.

1-Amido-7-naphthol-3-sulphonic acid

(NH_2, HO, S)

(**B-acid**) is obtained, mixed with the 1.3.7-acid, by heating α-naphthylamine-3-7-disulphonic acid with 40 per cent. caustic soda solution at 200° C. It forms sparingly soluble diazo salts, and it couples in acid or alkaline solution with diazo salts.

1-Amido-7-naphthol-4-sulphonic acid is obtained by sulphonation of 1-amido-7-naphthol with sulphuric acid below 30° C. It is similar in properties to the previous acid.

NH₂ OH

1-Amido-8-naphthol [structural formula] may be obtained by fusion of α-naphthylamine-8-sulphonic acid with caustic soda at 230° to 240° C., or by heating 1.8-diamido-naphthalene with sodium bisulphite solution or with dilute sulphuric acid solution under pressure.

1-Amido-8-naphthol-4-sulphonic acid (S-acid) may be obtained by heating 1.8-diamido-naphthalene-4-sulphonic acid with 40 per cent. sodium bisulphite solution at 95° C., in presence of acetone, and decomposition of the product with alkali.

It is used as middle component for disazo dyes and couples in either acid or alkaline solution.

OH NH₂

1-Amido-8-naphthol-6-sulphonic acid S[structural formula] is obtained from α-naphthylamine-6.8-disulphonic acid by heating with 50 per cent. caustic soda under pressure at 200° C. It forms sparingly soluble diazo salts and couples in alkaline solution with diazo salts.

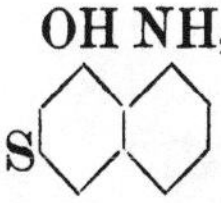

1-Amido-8-naphthol-3.6-disulphonic acid (H-

NH₂ OH

acid) S[structural formula]S may be obtained by heating 1.8-diamido-naphthalene-3.6-disulphonic acid with 10 per cent. sulphuric acid at 120° C., or with 40 per cent. caustic soda solution at 200° C.

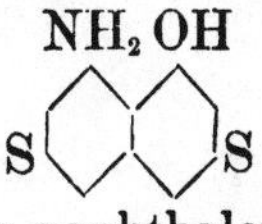

It forms soluble diazo salts, and also couples in acid

or alkaline solutions with diazo salts to form azo dyes. It is largely used for the production of azo dyestuffs.

2-Amido-1-naphthol (structure: OH, NH_2) may be obtained by reduction of 2-nitroso-1-naphthol. It has been used for the production of azo dyestuffs. It may be diazotised in neutral solutions in presence of copper salts to form naphthalene-2-diazo-1-oxide

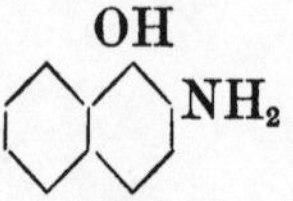

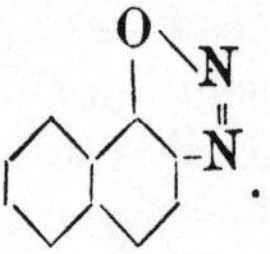

(structure: O, N, N.)

2-Amido-1-naphthol-4-sulphonic acid

(structure: OH, NH_2, S)

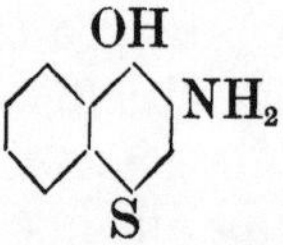

is obtained by reduction of nitroso-α-naphthol-4-sulphonic acid. It is diazotisable in neutral solutions giving a diazo oxide.

2-Amido-3-naphthol (structure: NH_2, OH) is obtained from the corresponding dioxy compound by heating with 30 per cent. ammonia at 140°—150° C., or with ammonium sulphite solution and ammonia at 80° C. It couples in alkaline solutions with diazo salts to form diazo compounds.

2-Amido-3-naphthol-6-sulphonic acid (R-acid) (structure: S, NH_2, OH) is obtained by heating β-naphthylamine-3.6-disulphonic acid with 75 per cent. caustic soda at 230° to 250° C. It is very sparingly soluble in water.

It may be diazotised and it couples in alkaline solution with diazo salts.

2-Amido-5-naphthol (NH_2, OH) may be obtained

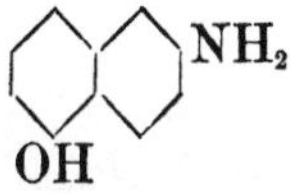

by heating β-naphthylamine-5-sulphonic acid with caustic soda at 260° to 270° C.

2-Amido-5-naphthol-7-sulphonic acid (J-acid)

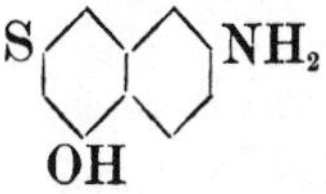

S, NH_2, OH is obtained from β-naphthylamine-5.7-

disulphonic acid by heating with caustic soda at 180° C., or with 50 per cent. caustic soda solution at 190° C. The β-naphthylamine-5.7-disulphonic acid is obtained by sulphonating β-naphthylamine-5-sulphonic acid with fuming sulphuric acid (20 per cent. anhydride) below 20° C. About a 50 per cent. yield of the 5.7-disulphonic acid is obtained.

It is very important because many dyestuffs obtained from it have good cotton-dyeing properties without the use of benzidine, its homologues, or sulphur bases (e.g., Primuline etc.).

N-ethyl and N-phenyl J-acids are also of great importance in the production of azo dyestuffs.

2-Amido-5-naphthol-1.7-disulphonic acid and the 3.7-disulphonic acid are also employed in the production of azo dyestuffs.

2-Amido-7-naphthol HO, NH_2 is obtained

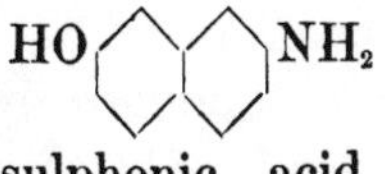

when β-naphthylamine-7-sulphonic acid is heated with a 50 per cent. solution of caustic soda at 260° to 300° C., or by boiling 2.7-diamido-naphthalene with sodium bisulphite solution followed by decomposition with caustic, or by heating 2.7-dioxynaphthalene with ammonium sulphite and ammonia under reflux condenser.

It forms diazo salts and it also couples in acid or alkaline solution with diazo salts to form azo compounds.

2 - Amido - 7 - naphthol - 3 - sulphonic acid ("F-acid") is obtained from 2.7-dioxynaphthalene-3-sulphonic acid by heating with ammonia ·880 sp. gr. at 120° to 150° C.

It forms sparingly soluble diazo salts and it couples in acid or alkaline solutions with diazo salts.

The 2-amido-7-naphthol-3.6-disulphonic acid forms sparingly soluble diazo salts and couples extremely slowly or not at all with diazo salts.

2-Amido-8-naphthol (OH, NH_2 [naphthalene ring structure]) may be obtained by heating β-naphthylamine-8-sulphonic acid with caustic soda at 260° to 270° C.

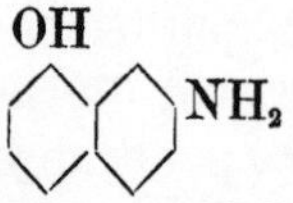

It couples in alkaline or in acetic acid solution with diazo salts to form azo compounds.

2-Amido-8-naphthol-6-sulphonic acid ("G- or γ-acid") is obtained from 2.8-dioxynaphthalene-6-sulphonic acid by heating with "880" ammonia at 120° to 150° C., or by heating with ammonia and ammonium sulphite at 150° C.

It is largely used in the production of azo dyestuffs.

2 - Amido - 8 - naphthol - 3 . 6 - disulphonic acid ("2R-acid") is obtained by heating β-naphthylamine-3.6.8-trisulphonic acid with 80 per cent. caustic soda at 220° to 250° C.

It forms sparingly soluble diazo salts and it couples in alkaline solution with diazo salts.

CHAPTER VIII

QUINONES

QUINONES are substances obtained from aromatic hydrocarbons by the replacement of two atoms of hydrogen by two atoms of oxygen. The simplest member is quinone or more correctly para-benzoquinone which is obtained by the oxidation of many benzene derivatives. Quinone is easily reduced to hydroquinone and hydroquinone is readily oxidised to quinone; quinone on treatment with phosphorus pentachloride gives hexachlorobenzene. From these simple reactions Graebe assigned to it a peroxide formula this being later replaced by the quinonoid formula since quinone gives a monoxime and this in turn a dioxime

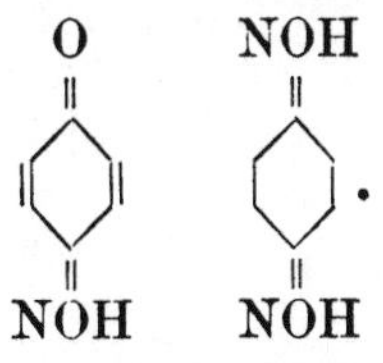

The monoxime is identical with para-nitrosophenol

which according to Sluiter exists as nitrosophenol (OH ⟨benzene ring⟩ NO) in the free state, whilst the salts are derived from the quinone-oxime.

Benzoquinones exist in two forms, the ortho and the para, the ortho again existing in two modifications, ⟨ring⟩(=O)(=O) a red coloured substance and ⟨ring⟩(—O—O—) a colourless body. By rapid oxidation of catechol in an inert solvent with silver oxide the colourless compound is obtained, but if the reaction takes place slowly the more stable bright red compound is formed. Small amounts of the colourless body are produced when the red body is recrystallised. The ortho-quinones condense fairly readily with ortho-diamines to form azines. Diphenazine being obtained thus

$$C_6H_4(=O)_2 + (H_2N-)_2C_6H_4 \rightarrow C_6H_4(-N=)_2C_6H_4 + 2H_2O$$

from ortho-benzoquinone and ortho-phenylene diamine.

Quinone derivatives are often formed as intermediate compounds during the preparation of dyestuffs. Thus the condensation of para-chlorophenol with a para-phenylene diamine or a para-amidophenol by means of oxidising agents gives ortho-quinone derivatives, e.g.,

$$Cl\,C_6H_3(=O)(=N-C_6H_4-N(CH_3)_2) \quad \text{or} \quad Cl-C_6H_3(=O)(=N-C_6H_4-OH).$$

Para-benzoquinone $O=C_6H_4=O$ is obtained by oxidation

of aniline with chromic acid. It is also obtained by the further oxidation of **aniline black** by means of chromic acid, many other azine dyestuffs also giving para-quinones on oxidation.

Para-benzoquinone forms yellow prisms, sublimes readily in gold-coloured needles, and possesses a peculiar penetrating odour. Quinone is fairly active and reacts with many compounds containing the amido group, e.g., ortho-nitraniline, anthranilic acid, diamido-diphenyl-methane, etc.

Tetrachloro-para-benzoquinone or **Chloranil**

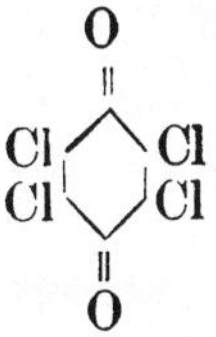

O
‖
Cl Cl
Cl Cl
‖
O

is produced by the action of potassium chlorate and hydrochloric acid on aniline; or from pentachloro-phenol by means of oxidising agents, the pentachloro-phenol being obtained by the action of hydrochloric acid and a chlorate on phenol; by the action of chlorate and hydrochloric acid on para-phenylene diamine.

Chloranil readily sublimes and melts in a sealed tube at 290° C. It is used in melt processes as an oxidising agent to convert leucobases into dyestuffs.

Naphthoquinones. There are six possible naphtho-quinones of which three are known and derivatives of a fourth are also known. They are coloured bodies and generally give coloured derivatives, a small number of these being commercial dyestuffs.

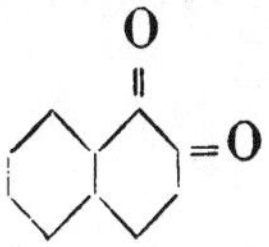

O
‖
=O

1 . 2-Naphthoquinone or **β-naphtho-**

quinone is often obtained when azo dyestuffs are reduced and the separated bodies are subsequently oxidised. It may be obtained by oxidising a dilute solution of 2-amido-α-naphthol in 5 per cent. sulphuric acid with a cold solution of potassium bichromate.

It crystallises in red needles and decomposes with blackening at 115° to 120° C. It is odourless and non-volatile in steam. Dilute caustic alkalis dissolve it, giving a yellow solution, which rapidly darkens on exposure to air.

When heated with acetic anhydride and sulphuric acid or zinc chloride, it gives a triacetate of 1.2.4-trihydroxy-naphthalene. And when heated with sulphuric acid at 120° to 150° C., it is converted into naphthoquinone black or **Naphthazarine.**

1 . 4-Naphthoquinone or **α-naphthoquinone** (M.P. 125° C.) may be obtained by the oxidation of naphthalene in acetic acid solution by chromic acid. It is also obtained by oxidation of simple α-or 1.4-disubstituted naphthalenes with chromic acid mixture.

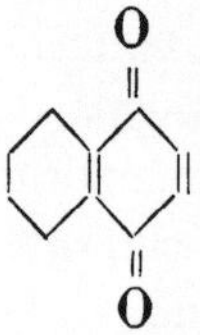

It crystallises in yellow needles, sublimes below 100° C., has an odour similar to para-benzoquinone, is volatile in steam, and dissolves in most organic solvents. In caustic soda solution it absorbs oxygen from the air forming 2-oxy-α-naphthoquinone.

2 . 6-Naphthoquinone 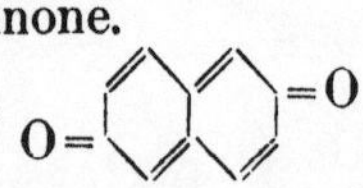 is obtained from the corresponding dioxy-compound by oxidation in boiling benzene solution with lead peroxide, it forms reddish yellow needles, is odourless, non-volatile, and is an oxidising agent similar to para-benzoquinone.

Naphthoquinone oximes may be obtained from the corresponding quinones by action of hydroxylamine, or by the action of nitrous acid upon naphthols or upon some of the oxynaphthoic acids. They are identical with the corresponding nitroso naphthols.

α-Naphthoquinone oxime or **4-nitroso-α-naphthol**

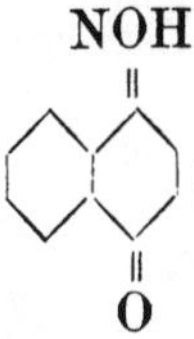

NOH

O

is not a dye, it does not form lakes. If an oxy group is introduced in the ortho-position to the oxime or keto group a dye is produced which will form lakes with coloured mordants. E.g. 2-oxy-α-naphthoquinone oxime

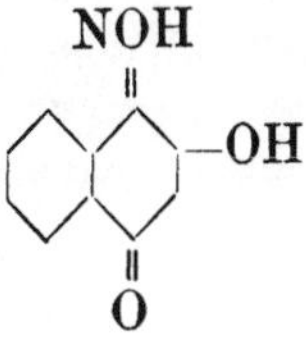

NOH –OH O

and 2-oxy-α-naphthoquinone

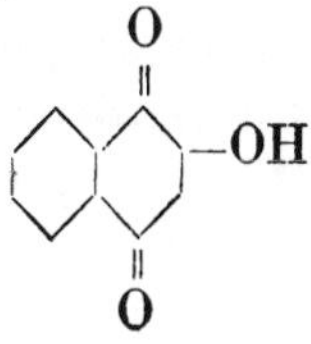

O –OH O

are dycs.

2-Nitroso-α-naphthol

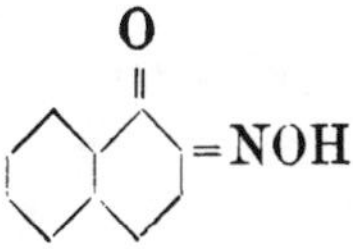

O =NOH

or **1-nitroso-β-naphthol**

NOH =O

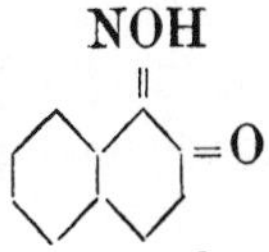

are dyes **Gambine R and V** (H) forming lakes, due, according to Tschugaeff and

Werner, to the oxygen of the carbonyl group having a residual valency (shown by dotted line) which forms a lake with a basic mordant

$$\begin{array}{l} -C=N-O\diagdown \\ \;\;| \qquad\qquad\quad Cr.OH. \\ -C=O\cdots\diagup \end{array}$$

Acenaphthenequinone $\begin{array}{c} CO-CO \\ |\quad\;\; | \end{array}$ (naphthalene rings) This body although termed a quinone is not strictly a quinone, the two carbonyl groups being situated in a 5-membered ring.

Chromic acid oxidises acenaphthene $\begin{array}{c} CH_2-CH_2 \\ |\quad\;\; | \end{array}$ (naphthalene rings) to acenaphthenequinone, the yield, however, is by no means good. It is prepared by adding 4 molecular proportions of amyl nitrite to a boiling solution of one molecular proportion of acenaphthene in alcohol, whilst a stream of hydrochloric acid gas is passed in. Two isomers are produced, one insoluble in sodium carbonate solution, which decomposes at 207° C., when rapidly crystallised from acetic acid. If the boiling with acetic acid is continued some time, the oxime (M.P. 230° C.) may be precipitated by addition of water. This oxime is dissolved in seven times its weight of 75 per cent. sulphuric acid and heated for one hour at 100° C.

Acenaphthenequinone condenses readily with substances containing an active methylene group. Vat dyestuffs may be produced in this way, e.g., **Ciba Scarlet G.**

The student seeking more detailed information on these bodies should consult the section on "Quinones" in Thorpe's *Dictionary of Applied Chemistry.*

Anthraquinone $C_6H_4{<}^{CO}_{CO}{>}C_6H_4$ (M.P. 277° C., B.P. 380° C.) is prepared from crude or purified anthracene, in a finely divided condition (which is obtained by rapidly quenching the distillate when anthracene is distilled in superheated steam), by boiling in a leaden-lined vat with sulphuric acid and sodium bichromate. The crude anthraquinone is filtered off, dried, dissolved in cold concentrated sulphuric acid, boiled with water, the precipitated anthraquinone filtered, boiled with soda solution, again filtered, washed and dried. The impurities are sulphonated and removed as salts. Another method is to mix the anthracene with copper oxide and heat at 250° to 300° C., air containing some nitrogen oxides is passed through the material, anthraquinone being produced. The electrolytic oxidation of anthracene, in an anode bath, with chromic acid solution as electrolyte, is practically of great importance; the crude anthraquinone being then purified as above; or the impurities are removed by solution in liquid sulphur dioxide, etc.

Anthraquinone is sparingly soluble in most organic solvents, it sublimes to form lemon yellow needles; it dissolves, without action, in sulphuric acid (98 per cent.) at 100° C. When strongly heated with sulphuric acid it is converted into mono- and di-sulphonic acids.

1.2-Anthraquinone [structural formula: three fused rings with =O groups at positions 1 and 2] and **1 . 4-**

anthraquinone (structural formula: anthracene skeleton with O= at 9 and 10 positions) have been obtained from β-anthrol (anthracene ring with –OH at β-position) and α-anthrol (anthracene ring with OH at α-position) respectively.

Anthraquinone-α-sulphonic acid $C_6H_4{<}(CO)_2{>}C_6H_3\cdot SO_3H$ (SO_3H in α-position).

During the sulphonation of anthraquinone the action of catalytic agents plays an important part, different catalytic agents controlling the position in which the sulphonic acid group enters. The catalytic agents generally used are boric and arsenic acids, and mercury salts.

The α-sulphonic acid is prepared by heating 100 parts of anthraquinone, 110 parts of sulphuric acid (30 per cent. SO_3), and 0·5 part of mercury to 130° C.

The free acid is readily soluble in water or alcohol; the calcium salt is more soluble in cold than in hot water and is crystallised from the hot solution.

Anthraquinone-β-sulphonic acid

$$C_6H_4{<}{-CO-\atop -CO-}{>}C_6H_3-SO_3H$$

One part of anthraquinone is mixed with one to one and a quarter parts of sulphuric acid (50 per cent. SO_3) and heated to 170° C., for about ten hours, the mixture being well stirred during the operation.

The product is diluted with water and filtered from

the unchanged anthraquinone, sodium carbonate is next added, which causes the sparingly soluble sodium salt of the β-acid to separate out, leaving the easily soluble sodium salt of the disulphonic acid in solution.

By recrystallisation of the mono-sodium sulphonate from water the so-called "silver salt" for alizarin manufacture is obtained.

Anthraquinone-disulphonic acids. By the further sulphonation of the α-acid in the presence of mercury the 1.5-anthraquinone disulphonic acid mixed with the 1.7- and the 1.8-acids are obtained. These acids are separated by the crystallisation of the calcium salts, the 1.8 is the least soluble, followed by the 1.5, whilst the 1.7 is easily soluble. The α-β-disulphonic acids of anthraquinone are also obtainable by sulphonation in two stages, ordinary and catalytic.

Phenanthraquinone

$$\begin{array}{l} C_6H_4-CO \\ \;\;\,| \qquad\quad | \\ C_6H_4-CO \end{array}$$

is obtained from the portion of coal tar boiling at 320° to 340° C., which is easily soluble in light petroleum or in 90 per cent. alcohol. The crude hydrocarbon, after removal of bases and acids, is oxidised with sulphuric acid and sodium bichromate solution; a vigorous reaction sets in, when this has subsided, the reaction mixture is heated to boiling for some time. The quinone is then precipitated with water, washed, filtered, dried, mixed with concentrated sulphuric acid and allowed to stand for one day. The phenanthraquinone, which is insoluble, is filtered off, washed with dilute alkali, and then extracted with warm strong sodium bisulphite solution in which the phenanthraquinone dissolves, and it is then precipitated from the filtered solution by sulphuric acid and bichromate.

Phenanthraquinone is a typical ortho-quinone, it gives the characteristic reactions of ortho-diketones; it condenses with ortho-diamines to give azines; it reacts with hot caustic alkali to give an oxy-acid, which on oxidation gives a ketone

$$\begin{array}{l} C_6H_4-CO \\ | \quad\;\; | \\ C_6H_4-CO \end{array} + KOH \rightarrow \begin{array}{l} C_6H_4 \\ | \end{array} \!\!\!\!> COH.COOK \rightarrow \begin{array}{l} C_6H_4 \\ | \\ C_6H_4 \end{array} \!\!\!\!> CO.$$

Phenanthrene resembles naphthalene more than anthracene; it may be nitrated similarly to naphthalene whereas anthracene is oxidised to anthraquinone.

Nitro-phenanthraquinones may be obtained from nitro-phenanthrenes or by nitration of phenanthraquinone.

CHAPTER IX

ACYLATION AND OXIDATION

Acylation.—By this term is meant the introduction of an acidyl group in place of hydrogen in hydroxy and amido compounds.

$$R.NH_2 + HOOC.R' = RNH.COR' + H_2O$$
$$R.OH + HOOC.R' = R.OOCR' + H_2O$$

The introduction of the acetyl group for example into aniline enables one on nitration to obtain almost pure para-nitraniline, whereas the nitration of aniline directly gives a large amount of tarry matter and a high proportion of meta-nitraniline.

In the oxidation of o-toluidine to anthranilic acid the acetylation of the amido group shields it against oxidation.

Acetylation of phenols is more difficult than that of amines; $POCl_3$ or PCl_3 may be used to remove water formed by the reaction.

The addition of a trace of pyridine assists acetylation.

In the production of some naphthylamine sulphonic acids the acetylation of the amido group shields certain positions in the naphthalene ring, e.g., 1.5-naphthylamine sulphonic acid. Without acetylation α-naphthylamine on direct sulphonation gives a mixture of 1.4 and other acids, with acetylation practically pure 1.5 acid is obtained.

The preparation of acyl derivatives of amido and oxy compounds plays an important part in the preparation of azo dyestuffs. Acylation modifies the properties of amines and phenols so that diazo bodies no longer condense with them. If a compound contains both amido and oxy groups, e.g., $\overset{NH_2\ OH}{S\ \bigcirc\!\!\bigcirc\ S}$, the oxy group may be shielded by acylation in alkaline solution, or in presence of dimethyl aniline by means of benzoyl-chloride; whereas the amido group may be acylated by means of concentrated acids or their anhydrides, generally at high temperature, or by the anhydride in neutral aqueous solution.

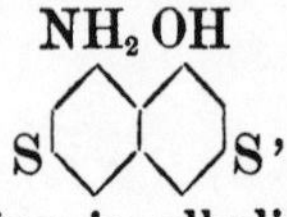

The most important acyl groups are formyl, acetyl, benzoyl and phthaloyl.

Alkylation and Arylation.—By these terms is understood the replacement of hydrogen by alkyl or aryl groups in amines, $R.NH_2 + R'Cl = RNHR + HCl$, phenols, $RONa + R'Cl = ROR' + NaCl$, and carboxylic acids. These reactions are applied both to intermediate products and dyestuffs. The alkylation of amines as a rule increases their rate of condensation with diazo compounds to form azo compounds. The same reaction applied to phenolic substances completely prevents coupling or condensation to form azo derivatives. The alkylation of amido groups in dyestuffs greatly affects the colour, e.g., rosaniline is converted into a spirit soluble blue by phenylation, or into a violet by methylation. Alkylation or arylation causes a change in colour towards the violet end of the spectrum. The introduction of the phenyl or benzyl group greatly aids sulphonation in converting basic dyes into acid dyes.

The shielding influence of aryl groups is of great importance with oxy and thioxy derivatives, resulting generally in increased fastness, especially towards washing and alkalis.

The most important alkyl groups are methyl, ethyl, benzyl and phenyl, and they may be introduced by means of the alkyl halide, by amines, or by alcohols in presence of hydrochloric or sulphuric acid.

Naphthols are easier to alkylate than benzene derivatives. Instead of using methyl chloride or dimethyl sulphate it is sufficient to heat with alcohol and sulphuric or hydrochloric acid or zinc chloride as dehydrating agents.

Arylation may be promoted by the use of other agents; see **Aniline Blue** (p. 195) for the use of benzoic or acetic acids. (Figs. III and VIII, Appendix.)

Oxidation.—A large number of intermediate compounds are obtained by the use of oxidising agents—the more important classes of compounds being aldehydes, ketones, acids and in some cases hydroxy compounds. Oxidation is also employed for the conversion of leuco compounds into dyestuffs.

For the preparation of these bodies the more important oxidising agents are air, chlorine, bichromates, permanganates, peroxides, e.g., lead, manganese, and hydrogen peroxides, persulphates, nitrites, nitrates, sulphuric acid particularly in presence of mercury salts (e.g., phthalic acid from naphthalene), hypochlorites, chlorates, ferricyanides, arsenates, and nitrobenzene and other nitro compounds.

CHAPTER X

ALDEHYDES AND CARBOXYLIC ACIDS

Aldehydes.

THE aromatic aldehydes and also formaldehyde are largely employed in the manufacture of triphenylmethane, acridine, and xanthene dyestuffs.

Formaldehyde H.CHO (B.P. – 21° C.) is obtained by passing a mixture of methyl alcohol vapour and air over a heated catalyst. Copper gauze in a copper tube is commonly employed. The tube is fitted with observation holes at each end, by means of which the approximate temperature may be gauged. The reaction mixture is passed through worms heated at about 40° to 45° C., in which the unattacked methyl alcohol is condensed, and may be rectified and again passed over the contact material. The formaldehyde passes on and is collected in water. The commercial article is a solution in water and is termed "**formalin.**" It contains from about 27 to 40 per cent. of formaldehyde. It is a pungent smelling gas. On keeping the aqueous solution it is slowly polymerised to "**paraform**" or trioxymethylene $(CH_2O)_3$, a crystalline solid of M.P. 152° C.

Formaldehyde condenses very readily with nearly all classes of aromatic compounds. Thus with aniline it forms anhydroformaldehyde aniline $C_6H_5N{=}CH_2$, a white solid, or in cold acid solution equimolecules combine to form para-amido-benzyl alcohol

$$H_2N-\langle\ \rangle-CH_2OH \text{ or } HN\langle\ \rangle-CH_2.$$

This body again readily condenses on heating with another molecule of aniline to form p.p-diamido-diphenyl methane.

In alkaline alcoholic solution two molecules of aniline condense with one molecule of formaldehyde to give methylene aniline $(C_6H_5NH)_2CH_2$.

Benzaldehyde C_6H_5CHO (B.P. 180° C., sp. gr. 1·05) is prepared by heating the mixture of benzal chloride and benzo-trichloride, obtained by the chlorination of toluene at the boil, with milk of lime under a pressure of from 4 to 5 atmospheres. Chalk is added along with the milk of lime in order to obtain a good emulsion. The benzaldehyde is distilled off with steam, whilst calcium benzoate, which is formed from the benzo-trichloride, remains behind, and is subsequently converted into benzoic acid by treatment with mineral acid.

$$C_6H_5CHCl_2 + Ca(OH)_2 = C_6H_5CHO + H_2O + CaCl_2$$

$$2C_6H_5CCl_3 + 4Ca(OH)_2 = (C_6H_5COO)_2Ca + 3CaCl_2 + 4H_2O$$

Benzaldehyde so obtained always contains some chlorine compounds. This renders it of little value for the scent industry, but it is quite suitable for the dye-stuff industry. Benzaldehyde free from chlorine is obtained by the oxidation of toluene in sulphuric acid solution by means of manganese dioxide.

Benzaldehyde is a colourless liquid smelling of bitter almonds. It is sparingly soluble in water. Commercial benzaldehyde should dissolve almost completely in a warm aqueous solution of sodium bisulphite of sp. gr. 1·11; on extraction with ether and evaporation of the ether there should be no pungent odour of benzyl chloride. It is employed in the manufacture of triphenylmethane dyestuffs.

Nitrobenzaldehydes.

Ortho-nitrobenzaldehyde $C_6H_4(NO_2)(CHO)$ (M.P. 43·5° C.)

may be obtained as follows:—

(1) By the oxidation of o-nitrobenzylaniline or its sulphonic acid, by means of sulphuric acid and either manganese dioxide or sodium bichromate, to the corresponding o-nitrobenzylidene derivative, which is then hydrolysed by dilute acid, and distilled with steam.

$$C_6H_4(CH_2Cl)(NO_2) + C_6H_5{-}NH_2 \xrightarrow{\text{condensation}} C_6H_4(NO_2){-}CH_2{-}NH{-}C_6H_5$$

$$\xrightarrow{\text{oxidation}} C_6H_4(NO_2){-}CH{=}N{-}C_6H_5 \xrightarrow{\text{hydrolysis}} C_6H_4(CHO)(NO_2) + C_6H_5{-}NH_2$$

(2) By the oxidation of o-nitrotoluene with sodium hypochlorite solution in presence of nickel oxide. It crystallises in long yellow needles, and is employed in the manufacture of indigo and triphenylmethane dyestuffs.

Meta-nitrobenzaldehyde $C_6H_4(NO_2)(CHO)$ (M.P. 58° C.)

is obtained by nitrating one volume of benzaldehyde with a mixture of five volumes of fuming nitric acid and ten volumes of concentrated sulphuric acid; the temperature must not exceed 10° C. The reaction mixture is poured into ice water.

It is a pale yellow compound, and is employed in the manufacture of triphenylmethane dyestuffs.

Para-nitrobenzaldehyde $C_6H_4(NO_2)(CHO)$ (M.P. 106° C.) is obtained by similar methods to those employed in the preparation of the ortho compound, or by boiling p-nitrobenzyl chloride with a saturated solution of cupric nitrate. Oxidation takes place with an almost theoretical yield of p-nitrobenzaldehyde. It crystallises from water in long thin prisms.

Chlorobenzaldehydes.

These compounds are also employed in the manufacture of the triphenylmethane dyestuffs.

Ortho-chlorobenzaldehyde $C_6H_4(Cl)(CHO)$ (B.P. 214° C., sp. gr. 1·29) is obtained by the oxidation of ortho-chlorotoluene, in sulphuric acid solution or suspension, by means of manganese dioxide; or by oxidation of benzyl sulphanilic acid, obtained by condensation of o-chlorobenzyl chloride with sulphanilic acid, to the benzylidene derivative, followed by hydrolysis (see o-nitrobenzaldehyde). It is a colourless liquid.

By heating with sodium sulphite under pressure it is converted into benzaldehyde-o-sulphonic acid, which is also obtained by oxidation of di-amido-stilbene-di-sulphonic acid (see p. 52).

Meta-chlorobenzaldehyde $C_6H_4(Cl)(CHO)$ (B.P. 212° C., sp. gr. 1·246) is obtained by the chlorination of benzaldehyde in presence of zinc chloride; or by diazotisation of m-amidobenzaldehyde and treatment of the diazo-chloride with metallic copper. It is a colourless oil.

Para-chlorobenzaldehyde $C_6H_4(Cl)(CHO)$ (M.P. 47·5° C., B.P. 213° to 214° C.) is often associated with the ortho compound, from which it may be separated by sulphonation with fuming sulphuric acid; or by the oxidation of p-chlorobenzyl aniline with sodium bichromate and sulphuric acid, followed by hydrolysis of the benzylidene compound with acid.

Dichlorobenzaldehydes are also employed in the manufacture of the triphenylmethane dyestuffs.

The 2.5-compound is obtained by the chlorination of benzaldehyde in presence of iodine or antimony. The crude compound boils at 231° to 233° C.

The 2.6-dichlorobenzaldehyde is obtained from the corresponding dichlorobenzal chloride.

Similarly 2.3.4- and 2.4.5-trichlorobenzal chlorides are converted into aldehydes when heated with fuming sulphuric acid at 100° C. The product is poured into ice water, the bisulphite compound is next obtained and decomposed by sodium carbonate.

Benzaldehyde-o-sulphonic acid is prepared by heating a strong solution of sodium sulphite with o-chlorobenzaldehyde under pressure at 200° C.

Also by the oxidation of stilbene disulphonic acid with sodium permanganate in the cold.

Para-nitrobenzaldehyde-o-sulphonic acid

$$C_6H_3(CHO)(SO_3H)(NO_2)$$

is prepared similarly to benzaldehyde-o-sulphonic acid by oxidising a cold solution of the sodium salt of dinitrostilbenedisulphonic acid with sodium permanganate.

Meta-amidobenzaldehyde $C_6H_4(CHO)(NH_2)$ is prepared from the sodium bisulphite compound of m-nitrobenzaldehyde by reduction with ferrous hydroxide. The compound readily polymerises and is therefore only obtained stable as salts.

Para-amidobenzaldehyde $C_6H_4(CHO)(NH_2)$ (M.P. 71° C.) is obtained by boiling p-nitrotoluene with an alkaline solution of sodium polysulphide. In this reaction the nitro group acts as an oxidising agent within the molecule of the substance.

Also by the reduction of p-nitrobenzylidene aniline with sodium polysulphide solution followed by acid hydrolysis.

Para-amidobenzaldehyde and its derivatives may also be prepared by heating phenylhydroxylamine or its derivatives with formalin, followed by hydrolysis of the anhydro compound which is produced by boiling with water. For the production of alkyl derivatives, the hydroxylamine derivative is condensed with formalin and a dialkylaniline.

p-dimethylamidobenzaldehyde (M.P. 73° C.) crystallises from hot water.

p-diethylamidobenzaldehyde (M.P. 41° C.) crystallises from water in plates.

It easily polymerises to a yellow insoluble compound.

Oxybenzaldehydes.

Of the three isomers only the meta and the para compounds are used for the preparation of dyestuffs. They are mainly employed for the production of triphenylmethane dyestuffs.

Meta-oxybenzaldehyde $C_6H_4(OH)(CHO)$ (M.P. 104° C., B.P. 240° C.) is obtained from the corresponding amido compound by heating its diazo compound with water. It crystallises from hot water in colourless needles.

Para-oxybenzaldehyde $C_6H_4(OH)(CHO)$ (M.P. 115° to 116° C.) may be obtained by the Tiemann-Reimer reaction of chloroform upon sodium phenate, in which case it is obtained along with salicylic aldehyde, the ortho and para compounds being separated by steam distillation.

The para compound is manufactured by the Gattermann method, which consists in the treatment of phenol with hydrocyanic and hydrochloric acids; there is obtained the chlor-imide which on treatment with water gives ammonium chloride and p-oxybenzaldehyde.

$$\text{C}_6\text{H}_5\text{OH} \rightarrow \text{HO}\cdot\text{C}_6\text{H}_4\cdot\text{HC}=\text{NH}\cdot\text{HCl} \rightarrow \text{HO}\cdot\text{C}_6\text{H}_4\cdot\text{CHO} + \text{NH}_4\text{Cl}.$$

Carboxylic Acids.

Benzoic acid C_6H_5COOH (M.P. 121° C., B.P. 249°C.). The major portion of this acid is obtained as a by-product in the manufacture of benzaldehyde. During the preparation of benzal chloride some benzotrichloride is produced, which on hydrolysis with milk of lime gives calcium benzoate, from which benzoic acid is obtained by acidifying with hydrochloric acid.

Benzoic acid may be obtained by boiling one volume of benzyl chloride with three volumes of dilute nitric acid (16 per cent.) until the liquid no longer smells of benzyl chloride or benzaldehyde. Also by heating benzotrichloride with acetic acid and zinc chloride, the acetyl chloride obtained as a by-product being distilled off.

$$C_6H_5CCl_3 + 2CH_3COOH \rightarrow C_6H_5COOH + 2CH_3COCl + HCl.$$

The commercial benzoic acid usually contains some chlorine derivatives; the quantity should only be small, in which case it is not detrimental for general use.

It is employed for the production of benzoyl chloride and as an assistant in the arylation of amido compounds.

Benzoyl chloride C_6H_5COCl (B.P. 198·5° C., sp. gr. 1·23) may be prepared by heating benzoic acid with phosphorus pentachloride, the phosphorus oxychloride is distilled off and the benzoyl chloride fractionated.

Benzoyl chloride is manufactured from calcium benzoate by treating with chlorine and sulphur dioxide.

It is a liquid of unpleasant odour, and is employed for the benzoylation of hydroxy and amido groups.

Salicylic acid $C_6H_4(OH)(COOH)$ (M.P. 156·7° C.). Phenol is dissolved in the equivalent quantity of concentrated caustic soda and the solution evaporated to dryness. The residue is powdered and dried in a current of an inert gas, the dried sodium phenate is saturated under pressure with dry carbon dioxide, sodium phenyl carbonate is produced, which on heating at 120° to 140° C. at ten atmospheres pressure is converted quantitatively into sodium salicylate, which crystallises in plates from water.

$$C_6H_5{-}ONa + CO_2 \rightarrow C_6H_5{-}CO_3Na \rightarrow C_6H_4(OH)(COONa)$$

It is employed as the azo component in many mordant azo dyes, and in the production of some cotton dyestuffs. It is also used in the preparation of "**salol**" (phenyl salicylate), and of methyl salicylate (oil of wintergreen), which is used as a disinfectant.

Cresotinic acid $C_6H_3(CH_3)(OH)(COOH)$ (M.P. 163° to 164° C.) is prepared similarly to salicylic acid from ortho-cresol. It is volatile in steam.

Nitrosalicylic acid $C_6H_3(OH)(COOH)(NO_2)$ (M.P. 230° C.) is obtained by nitration of salicylic acid with a mixture of nitric and sulphuric acids. The nitro acid is precipitated by pouring into water.

Amidosalicylic acid is obtained from the above compound by reduction with cold sodium sulphide solution or by zinc dust and acid.

Meta-oxybenzoic acid $C_6H_4(OH)(COOH)$ [ring –OH, COOH] (M.P. 188° C.) may be prepared by the action of nitrous acid upon m-amidobenzoic acid, or by heating m-chlorobenzoic acid, or m-cresol with sodium hydroxide under pressure. By heating with sulphuric acid (90 per cent.) at 210° C., anthraflavin and anthrarufin are produced.

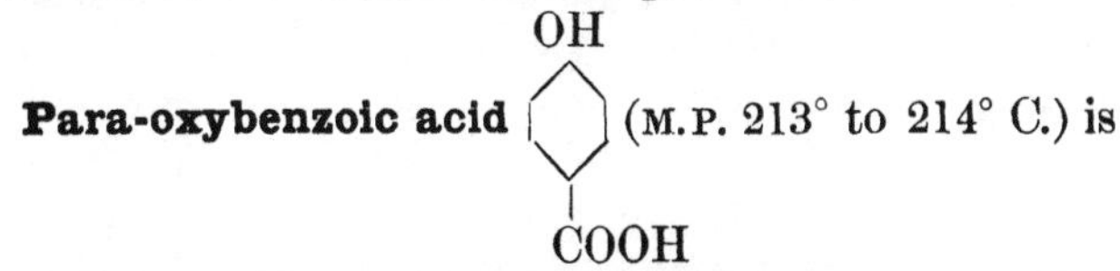

Para-oxybenzoic acid (M.P. 213° to 214° C.) is manufactured by the same method as the ortho acid except that during the transformation of the sodium phenyl carbonate the temperature of the reaction is 220° C. The acid may also be obtained by heating sodium salicylate. It crystallises from water with one molecule of water of crystallisation.

Oxynaphthoic acids. These acids are very interesting on account of the ease with which some of them lose carbon dioxide.

The **1-oxy-2-naphthoic acid** (M.P. 187° C.) is obtained from dry sodium α-naphtholate and carbon dioxide under pressure at 120° to 145° C. The acid crystallises in needles. Nitrous acid reacts with it, with formation of 2-nitroso-α-naphthol and elimination of carbon dioxide. It couples with diazotised amines forming para-azo dyes.

2-oxy-1-naphthoic acid is obtained by heating dry sodium β-naphtholate at 120° to 140° C., after saturating with carbon dioxide under pressure. The acid when quickly heated melts at 156° to 157° C., but if slowly heated begins to decompose between 124° and 128° C. It is converted by prolonged boiling with water into β-naphthol and carbon dioxide. When treated with diazo salts carbon dioxide is split off giving azo dyes of β-naphthol.

2-oxy-3-naphthoic acid (M.P. 216° C.) is obtained by heating dry sodium β-naphtholate with the required quantity of carbon dioxide at 200° to 250° C. under pressure. It crystallises in yellow scales, and is characterised by its greater stability, being decomposed, with loss of carbon dioxide, only when boiled with sodium bisulphite solution. Heated with ammonia under pressure at 260° to 280° C., it gives 2-amino-3-naphthoic acid. Nitrous acid converts it into 1-nitroso-2-oxy-3-naphthoic acid. It couples with diazo salts to give azo dyes without elimination of the carboxylic acid group.

The anilides of oxy-naphthoic acid are employed, in place of β-naphthol, for coupling on the fibre in the production of insoluble azo dyes. (Compare **Para-Red.**) They are put on the market as **Naphthol AS** and **NA** (G.E.).

CHAPTER XI

SULPHUR COMPOUNDS

THIO compounds are not used to a large extent in the production of dyestuffs, except perhaps in some of the vat dyes, many of these dyes themselves being thio compounds. The thio derivatives are, generally speaking, stable bodies and many of them give colours that are fairly fast to light, etc.

Many of the thio derivatives are extremely dangerous substances on account of their action upon the skin, causing eczema and blood poisoning. For this reason many of these compounds are converted into dyestuffs without purification.

Thioaniline H_2N –S– NH_2 (M.P. 105° C.) is employed in the production of a few azo dyes. It is obtained by heating aniline with sulphur at 150° to 160° C., until sulphuretted hydrogen is no longer given off. The excess of aniline is then removed by steam distillation and the thioaniline separated as sulphate by addition of sulphuric acid.

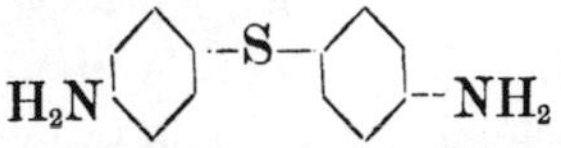

Dehydrothio-paratoluidine

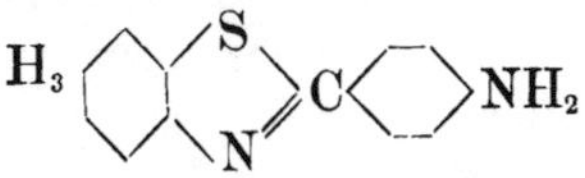

H_3 S C NH_2 N

(M.P. 191° C., B.P. 434° C.) is obtained, together with **Primuline** base, by heating paratoluidine (seven parts)

with sulphur (four parts) at 180° to 250° C. until sulphuretted hydrogen is no longer evolved. It is then separated from the primuline base by extraction with solvents.

It is a yellow crystalline body, sparingly soluble in water, fairly soluble in alcohol. Its salts are decomposed by water, and its diazo salts are easily soluble.

The sulphonic acid is prepared from the above base by heating at 40° to 50° C. with fuming sulphuric acid (50 per cent. SO_3). It crystallises in needles or leaflets, almost insoluble in water.

Dehydrothioxylidine H_3C—(benzene ring, with CH_3)—$\langle{}^{S}_{N}\rangle$C—(benzene ring, with CH_3)—NH_2

(M.P. 107° C., B.P. 283° C. at 14 mm.) is prepared similarly to the above base, the temperature being maintained at 185° to 190° C., until sulphuretted hydrogen is no longer evolved. The product is distilled in vacuo and extracted with hydrochloric acid (30 per cent.) in which the base is soluble, whereas isodehydrothioxylidine, formed simultaneously, is insoluble in hydrochloric acid.

It forms yellowish-white prisms.

The **isodehydrothioxylidine**

H_3C—(benzene ring, with CH_3)—$\langle{}^{S}_{N}\rangle$C—(benzene ring, with NH_2 and CH_3)

(M.P. 121° C.) forms yellow needles.

CHAPTER XII

DIAZO COMPOUNDS

THE diazo compounds are obtained by the action of nitrous acid upon salts of primary aromatic amines. In most cases the amine is dissolved in excess of mineral acid, generally hydrochloric or sulphuric, and sodium nitrite, powdered or in solution, is added in equivalent amount to the amine. The temperature at which the reaction is carried out is varied according to the properties of the compound treated. The method of formation of the diazo compounds shows that the nitrogen directly attached to the ring is pentavalent and the nitrogen introduced trivalent.

$$\underset{\text{Aniline hydrochloride}}{C_6H_5.NH_2.HCl} + \underset{\text{nitrous acid}}{HONO} \rightarrow \begin{array}{l} C_6H_5\text{—}N\text{—}Cl \\ \qquad\;\;\;\, ||| \\ \qquad\;\;\;\, N \end{array} + 2H_2O$$

Compounds of this structure are termed diazonium salts. These salts may undergo transformation to form the diazo salts, the latter existing in two stereo-isomeric modifications, viz.

$$\underset{\text{Syn-diazobenzene chloride}}{\begin{array}{c} C_6H_5\text{—}N \\ \quad\; || \\ Cl\text{—}N \end{array}} \qquad \underset{\text{Anti-diazobenzene chloride}}{\begin{array}{c} C_6H_5\text{—}N \\ \qquad || \\ \qquad N\text{—}Cl \end{array}}$$

The syn-diazo compounds readily couple with phenols or aromatic tertiary amines to produce azo compounds.

By the addition of a large excess of alkali to the syn compound it is transformed into the more stable anti-compound, which either does not condense with phenols or amines to produce azo compounds, or condenses only slowly. If hydrochloric acid is added to the anti-form it is transformed into the diazonium form, and this by the addition of a small amount of alkali gives the labile syn form (Hantzsch).

$$
\begin{array}{ccc}
 & C_6H_5—N—Cl & \\
 & \overset{|||}{N} & \\
\swarrow^{NaOH} & & \nwarrow^{HCl} \\
C_6H_5—N & \xrightarrow[\text{alkali}]{\text{excess}} & C_6H_5—N \\
\| & & \| \\
Cl—N & & N—Cl \\
\text{Syn} & & \text{Anti}
\end{array}
$$

The rate of isomerisation of syn- to anti- diazo compounds depends largely upon the groups present in the aromatic nucleus. Methyl groups hinder the rate of change, whilst halogen substituents increase it.

Diazophenols. The diazo chlorides of o- and p-amidophenol are obtained by diazotising the corresponding bases in alcoholic solution with amylnitrite and hydrochloric acid at 0° C., and precipitating with ether. When these salts are dissolved in water and treated with caustic potash, hydrochloric acid is split off, and the free diazophenols, or quinone-diazides, are formed.

$$C_6H_4\begin{matrix}—OH \\ —N:N.Cl\end{matrix} \rightarrow HCl + C_6H_4\begin{matrix}—O \\ \quad \diagdown \\ —N:N\end{matrix}$$

$$\text{or } C_6H_4\begin{matrix}=O \\ =N:N\end{matrix} \text{ or } C_6H_4\begin{matrix}=O \\ —N \\ \diagdown \| \\ N\end{matrix}$$

Meta-diazophenol chloride is very unstable, and loses nitrogen at 0° C.

By the diazotisation of substituted amidophenols quinone-diazides are obtained directly.

2.4.6-trichlorodiazobenzene acid sulphate on standing for some time loses one atom of chlorine and becomes converted into 3.5-dichloro-o-quinone-diazide.

On first consideration it appears a comparatively simple matter to produce diazo salts, nevertheless there are many amines which form diazo compounds very slowly, or resist the action of nitrous acid, or, owing to the formation of secondary products, are incapable of forming diazo salts.

For the diazotisation of amines the general method is to dissolve the amine in 2½ to 3 equivalents of hydrochloric acid, cool to about 5° C. and add the calculated quantity of sodium nitrite; in most cases slowly until, after standing for about five minutes, a reaction is obtained with starch-iodide paper. In certain cases, to prevent the formation of diazoamido derivatives, the reacting quantity of nitrite is added at once, care being taken that the temperature does not rise too high.

Symmetrical trinitroaniline is very difficult to diazotise unless a large excess of sulphuric acid is used as solvent. The use of strong acid solutions will often overcome the steric hindrance of various groups in substituted benzenes.

The ortho-amidophenols may be easily diazotised, but difficulties arise with the 1.2- and 2.1-amidonaphthols, owing to the oxidising action of the nitrous acid. This may often be overcome by the addition of salts of copper, zinc, nickel and mercury to the reaction mixture.

By the action of nitrous acid upon aromatic compounds containing more than one amido group, one would expect each amido group to be converted into the corresponding diazo group. In many cases secondary reactions take place with such rapidity that no diazo salt can be isolated, and special methods are employed for these compounds.

For example, orthophenylene diamine gives, on treatment with sodium nitrite and a mineral acid, aziminobenzene. Metaphenylene diamine similarly treated gives the well-known dyestuff **Bismarck Brown.** By the rapid addition of metaphenylene diamine hydrochloride to an excess of nitrous acid a clear yellow solution of the tetrazo compound is obtained. The substituted metaphenylene diamines, particularly 2.6-diamido-4-chlorophenol, 2.6-diamido-p-cresol, 3.5-diamido-p-oxybenzoic acid, and 2.4-diamidophenol, readily tetrazotise. Only one amido group is diazotised if the theoretical amount of nitrite is mixed with the base and mineral acid then added.

Paraphenylene diamine on treatment with one equivalent of sodium nitrite and mineral acid gives a mixture of diazo- and tetrazo-compounds. The diazo compound is obtained indirectly, using p-amido-acetanilide as base, or by employing p-nitraniline. The diazo compound is prepared and, after coupling to form the azo compound, the nitro group is reduced by sodium sulphide solution, or the acetyl group is removed by hydrolysis with caustic soda.

Nitro-para-phenylene diamine hydrochloride by treatment with excess of acetic acid and excess of sodium nitrite gives only the diazo compound, which is probably 3-nitro-4-amido-diazobenzene chloride. If, however, this diazo compound is coupled with a component, then the free amido group may be diazotised.

The benzidine derivatives may be either diazotised or tetrazotised.

The ortho- and the peri-diamido derivatives of naphthalene react with nitrous acid to give azimino compounds. With 1 . 4-naphthalene diamine it is necessary to shield one amido group, by acylation, diazotise, couple, hydrolyse, and again diazotise. This is necessary to prevent oxidation with formation of naphthoquinone. With some of the sulphonic acid derivatives of 1 . 4-naphthalene diamine one amido group only is attacked when treated with nitrite and acetic or oxalic acid; after coupling the other amido group may be diazotised.

Sulphur dioxide or sulphites react with diazo compounds to produce series of compounds depending upon the conditions. Sulphazides, diazobenzene sulphones, sulphinic acids, or diazo-sulphites may be produced. The diazo-sulphites are important as intermediate products for the preparation of hydrazine and its derivatives.

The sulphides and disulphides are produced from diazo salts by interaction with sulphuretted hydrogen in nearly neutral or in acid solution. The disulphides are important intermediate products for the preparation of mercaptans, the latter being employed in the production of many vat dyestuffs.

Thiosalicylic acid $C_6H_4\begin{matrix}-SH\\-COOH\end{matrix}$ is obtained thus:— Anthranilic acid (1 mol.) is diazotised with the theoretical quantity of sodium nitrite (1 mol.) in hydrochloric acid solution, the temperature being kept below 5° C.

This solution is now run into an ice-cold solution of sulphur (1 atom) dissolved in sodium sulphide (3 mols.) and caustic soda (1 mol.) of 76° Tw. During the addition of the diazo solution ice is added to keep the temperature below 5° C. Nitrogen is now rapidly given off, the temperature rising quickly; after allowing to stand some hours the solution is acidified with hydrochloric acid, using **Congo Red** paper as indicator. **Dithiosalicylic acid** $C_6H_4(COOH)-S-S-C_6H_4(COOH)$ separates, it is filtered off and washed with water. The precipitate is dissolved in a solution of sodium carbonate, filtered from sulphur, and, after the addition of iron or zinc dust ($1\frac{1}{2}$ mols.), heated to boiling until sulphuretted hydrogen is no longer recognisable on acidifying and boiling. The iron or zinc is then precipitated with caustic soda, boiled and filtered. The thiosalicylic acid is then precipitated from the filtrate by addition of sulphuric acid.

Some curious reactions have been observed when certain dinitro derivatives of aniline are diazotised. When 3.4-dinitro-o-anisidine is diazotised in acetic acid solution, one of the nitro groups is eliminated during the operation, and is replaced by the hydroxyl group, the compound formed having the composition

$$N_2{=}C_6H_2(OCH_3)(NO_2){=}O$$

The diazotisation of 2.3-dinitro-p-anisidine in acetic acid solution also causes one nitro group to be replaced by an oxy group. In nitric or sulphuric acid solution the nitro group remains unaffected, but in presence of hydrochloric acid the nitro group adjacent to the diazo group is replaced by chlorine.

It has also been observed that when a methoxy group is in the para position to the amido group, and at the same time has a nitro group in an adjacent position, demethylation takes place on diazotisation. Thus 3.5- and 2.5-dinitro-p-anisidines yield the corresponding quinonediazides of dinitrobenzene.

With derivatives of napththylamines the number of these peculiar reactions is higher. Chlorine, bromine, and sulphonic acid groups are also readily replaced by hydroxyl.

Diazo derivatives of amido-anthraquinones have not yet been employed for the manufacture of dyestuffs, but their diazo compounds may be obtained by diazotisation in concentrated sulphuric acid solution or suspension in the same.

For a detailed treatment of this subject see Cain's *Chemistry of Diazo Compounds* (1907).

To the class of intermediate products belong also many of the explosives, disinfectants, synthetic drugs, perfumes etc. manufactured by the dyestuff firms. Separate sections cannot be devoted to these classes in a text-book such as this. Reference may however be made to Thorpe's *Dictionary of Applied Chemistry*, to literature specially devoted to these subjects and articles in technical journals.

PART II

DYESTUFFS

CHAPTER XIII

APPLICATION OF DYESTUFFS

FOR convenience in application dyes are grouped according to properties which are common to certain dyes causing them to be applied by similar methods. Thus the dyer finds it convenient to group dyes as acid, basic, mordant, direct cotton, vat or sulphide dyes without regard to the chromophore groups or the relations between colour and constitution.

The chief use of colouring matters is for the decoration of textile fabrics and the chief methods of application may be briefly stated in reference to the main uses.

Acid Dyes. This class is composed entirely of coal-tar dyestuffs, except for the natural dyestuff, **cudbear** and **orchil.** They are mostly sodium salts of coloured organic acids, e.g., sulphonic acids, carboxylic acids and nitrophenols. They are put on the market as powders containing sodium sulphate or chloride and often sodium carbonate. The system of diluting to type with these materials in order to maintain a standard article on the market easily lends itself to adulteration for the purpose of producing an apparently cheaper dye powder. The valuation of a dye sample may be accomplished by comparative experimental dyeing under conditions similar to those employed in practice. Where the constitution is known many dyes may be estimated by means of titanous chloride (Knecht). Spectroscopic analysis of dyes for purity, mixtures, etc. according to the method of Formanek is in restricted use owing to limited data and the expensive apparatus required. Different brands of the same dye are often obtained by grinding in a small amount of a shading-off colour to modify the shade for trade purposes as a distinct product. This may be extended to the production of mixtures of primary colours, etc. to give any desired shade. Many dyes for union materials are mixtures of this type, e.g., acid dyes dyeing well from neutral baths and direct cotton dyes. Such a mixture will dye a wool and cotton union in a single dyebath. The detection of such mechanical mixtures is done by blowing a puff of the dye powder on to a filter paper wetted with water, alcohol, etc. and examining the colour of the dissolving specks of powder. Or use is made of differences in diffusibility by dipping filter paper in a solution of the dye and examining different zones of colour formed above the surface by capillary attraction (Goppelsroeder).

More rarely acid dyes (so-called) are found containing metallic salts, as for instance those of chromium, copper, magnesium, zinc, etc., where these produce some desired modification in properties, as in the case of acid-mordant dyes.

The application of acid dyes to animal fibres (wool, mohair, etc., silk and wild silks) is from hot acid baths. **Example.** Medium shades of colours are obtained on wool with 2 per cent. of its weight of dye, 2 per cent. of sulphuric acid and 10 per cent. sodium sulphate in a bath equal to thirty times the weight of wool. The acid aids dyeing and may be replaced by formic acid, acetic acid or ammonium acetate where there is a liability to dye unevenly or to change shade. The sodium sulphate assists in the production of level shades and is increased where a tendency to unevenness is to be counteracted. About 5 per cent. dye is required to dye blacks. Certain acid dyes are applied according to special methods.

Acid dyes have no affinity for cotton but may be applied by padding and drying the dye on the fibre, giving shades fast only to light.

The fastness of acid dyes on wool varies, but on the whole is good to common agencies, e.g., light and washing, and these dyes are considerably used on dress materials, carpet yarn, and various unions containing wool, also on silk fabrics.

Basic Dyes. This class contains the brightest and strongest colouring matters. They are mostly hydrochlorides of coloured bases, e.g., **Magenta, Methyl Violet,** etc. The only well-known natural dye of the basic class is **berberine** (barberry), now hardly ever used. The purer types of certain basic dyes appear on the market as crystals, e.g., **Diamond Magenta, Malachite Green crystals.** Dextrin is sometimes present as an adulterant in powders. The basic dyes dye animal fibres direct from aqueous solution. Restraint on the dyeing is secured by a slight addition of acid. **Example.** A medium shade of most basic dyes is obtained on wool by applying 1 per cent. dye with ½ per cent. acetic acid from a bath 30 times the weight of the wool being dyed, with ½ hour's heating to 95° C. and ½—1 hour's heating at 95° C. A medium shade on cotton is obtained by first mordanting the cotton in a cooling bath of tannin (3 per cent.) followed by fixing the tannin with a cold bath of tartar emetic (1½ per cent.). **Auramine** is dyed not higher than 80° C. as it decomposes at temperatures near the boil. Acid mordants other than tannin may also be used, e.g., oil mordants. The fastness of basic dyes is inferior to that of acid dyes, and on that account their use is restricted mainly to the production of bright shades unobtainable with other classes of dyes, and to the "shading-off" or "topping" of shades dyed by other methods.

Direct Cotton Dyes, Substantive or Salt Dyes. These dyes are acid colours having a special affinity for cotton, except a small class of strongly basic dyes, e.g. the **Janus** dyes, which dye cotton direct. Most of them are derivatives of benzidine, but many other products are known which will dye unmordanted cotton. The only well-known natural dye of this class is **turmeric**, which is no longer used to any large extent for dyeing. They are sold as powders, usually containing sodium sulphate or chloride and sodium carbonate. When dyed from a neutral or slightly alkaline bath the dye is taken up as a whole, i.e. the sodium as well as the colour acid, hence the name Salt Dyes. They are mainly used for cotton and cellulose fibres, but certain of these dyes also find application to animal fibres, in which case dyeing may be done from a slightly acid bath. **Example.** For a medium shade 2 per cent. **Benzopurpurine** may be dyed on cotton from a bath twenty times the weight of the material, containing 10 per cent. sodium chloride or sulphate and 1 per cent. sodium carbonate. The cotton may be entered cold, raised to near the boil in half an hour and maintained for half an hour at this temperature.

The fastness of these dyes is generally good to usual agencies, but in many cases may be considerably improved by a suitable treatment after dyeing (see page 160). They are used for all ordinary purposes on cotton, in calico printing, on linen, ramie, paper, wool and silk. The natural direct dyes **turmeric** and **annatto** are fugitive.

Mordant Dyes. These colouring matters possess a special acid character enabling them to form insoluble lakes with metallic mordants, such as salts of tin, aluminium, chromium, copper, iron, etc. The colour of the lakes given by any one dyestuff and different mordants may vary widely, tin giving the brightest shades and iron the dullest. Fastness to various agencies also varies with the mordant used, e.g., iron and copper are noted for giving shades fast to light, but shades on iron mordants are liable to be not so fast to acids as on chrome mordants, which again are noted for fastness to washing and are mostly used on wool. The bright shades on tin and alum mordants are not nearly so fast to light or washing as the shades produced by use of chrome.

The substituent groups giving mordant dyeing properties are OH, COOH and NO. According to rule (Liebermann and Kostanecki) two of these groups must be present in a dyestuff, ortho to each other, to give mordant dyeing properties. There are, however, many exceptions to such a statement, although mordant dyes in general comply with it. In the azo series dyes having mordant dyeing properties are found containing only one OH group, ortho to the azo group. It will be seen that dyes complying with other conditions of grouping and application may also in addition possess mordant properties, thus there are Acid Mordant Dyes and Salt Mordant Dyes, which are faster when fixed with chrome than when merely dyed as Acid or Salt Dyes. At the same time chroming always affects the shade more or less. Such products are soluble in water, whereas the mordant dye *per se* is rarely very soluble. The latter are put out as 20 per cent. pastes, dyeing being done with the colouring matter in suspension. Water soluble (s.w.) brands of these dyes are often produced by the action of bisulphite on ketonic dyes, e.g. alizarines. With these products care is taken to control the temperature during dyeing so as to avoid rapid decomposition in the bath. **Example.** Wool is boiled for 1½—2 hours with 3 per cent. bichrome ($K_2Cr_2O_7$ or $Na_2Cr_2O_7$, $2H_2O$) and 2½ per cent. tartar in a bath thirty times the weight of the wool. It may then be dyed similarly with 5 per cent. **Anthracene Brown** paste, in a bath of similar volume containing 1 per cent. acetic acid.

Most dyes of this class are *polygenetic*, i.e., give different shades with different metallic mordants.

Cotton may be mordanted with Turkey Red Oil by steeping and drying when it will then take up alumina. Or acetates of alumina, chrome, or iron, may be steamed on the cloth and the combined acetic acid driven off, the alumina or other hydrate being then combined with oil. Acid mordant dyes may be first dyed on wool like acid dyes and then after-chromed. In some cases chrome compounds may be present from the start, e.g. **Metachrome** dyes.

The chief natural dyes in use are of the mordant class. The good properties of certain yellow dyes such as **fustic, weld, quercitron bark** and **Persian berries** have prevented their entire disuse.

The **redwoods, cochineal,** and **logwood** are other natural mordant dyes in extensive use. Ready prepared extracts of these products are often used for dyeing, e.g. **Flavin, Fustin, Ammoniacal Cochineal, Logwood Extract** and **Haematein Crystals.** The methods of dyeing with these products are not essentially different from the methods of applying artificial mordant dyes. **Logwood** or **Haematein** is dyed on a chrome, iron or copper mordant, each mordant being characterised by different properties of resultant shade and fastness, these being functions of the nature of the lake formed on the fibre. **Cochineal** is usually dyed with a tin mordant in the dyebath, oxalic acid being added to prevent lake formation in the bath. Other dyes are first dyed and then "saddened" with a mordanting salt in a similar way to the after-chroming of an acid-mordant dye, after first dyeing on wool as an acid dye.

Vat Dyes. **Indigo** is the chief dye of this class, and until comparatively recently was the only dye of the type in use. The vat dyes are stable insoluble colouring matters which by reduction, as for example with hydrosulphite of soda, zinc dust, glucose, etc. along with alkalies, give solutions of their leuco compounds from which they may be dyed on cotton and also wool and silk, if the alkalinity of the dyebath is not required to be such as would injure animal fibres.

Indigo is also dyed from fermentation vats, reduction being effected by the aid of micro-organisms. The true shade is not developed in vat dyeing until the material is removed from the vat liquor, and the leuco body taken up by the fibres re-oxidised to the colouring matter, either by air oxidation or an oxidising agent. The stability and insolubility of vat dyes also afford some measure of their permanency when applied on textile materials, but it should be remembered that a vat dyestuff need not inevitably be fast, and although the highest degree of fastness is attainable by the use of dyes of this class there are others which are by no means fitted for production of the fastest shades.

The quantities of hydrosulphite and caustic soda required vary with different dyes as does also the temperature of dyeing.

Sulphide Dyes. These are dyed in the reduced state from alkaline baths containing sodium sulphide which acts both as reducing agent and alkali. Sodium carbonate is frequently added also to increase the alkalinity and common salt is added to aid the dyeing. Owing to the alkalinity of the bath the use of sulphide dyes is almost entirely restricted to vegetable fibres which are usually dyed at or near the boil. The full shade only develops by subsequent oxidation with air or oxidising agents, of which bichrome and peroxide are most in use.

Dyes produced on the Fibre. This class includes **Aniline Black, Para-Red, Nitroso-Blue** and a few **Mineral Dyes,** such as **Iron Buff, Prussian Blue, Chrome Yellow** and **Manganese Bronze.** These latter are produced on the cotton fibre by successive paddings or steeping, followed by squeezing-off and immersing in a reagent to develop the mineral colour. For example, if cotton is steeped in basic lead acetate and then in bichromate of potash, insoluble lead chromate or **Chrome Yellow** is formed on the fibre. It is not suitable for wool owing to the sulphur in the fibre. Mere steeping in a solution of permanganate causes the deposition of brown manganese hydroxide on cotton or animal fibres, e.g., furs, the tips of which may be discharged white by brushing with bisulphite of soda. **Prussian Blue** is produced on wool by heating in a bath of potassium ferrocyanide and sulphuric acid.

The dyeing of silk is often done in a bath containing "boiled-off liquor" kept over from the removal of gum from raw silk with soap.

The dyeing of leather is on similar principles to the dyeing of wool, acid and basic dyes being mainly used. Certain basic dyes find their main use for this purpose, e.g., **Phosphine.**

The dyeing of skins and furs is done after degreasing. Acid or basic dyes are used and also colours produced by oxidation of amidophenols or diamines, e.g., with peroxide, these latter giving brown and black shades. Feathers are usually dyed with acid dyestuffs in an acid bath.

As regards vegetable fibres, linen is dyed like cotton, but penetration is slower. The same applies to ramie (rhea) or China grass. Hemp is mainly dyed with basic dyes. Fibres containing lignin dye full shades with basic dyes without a previous mordanting with tannin, e.g., jute, cocoanut fibre. Straw and wood chip are commonly dyed with basic or acid dyes, but also more and more with direct cotton dyes in boiling baths containing Glauber's salt. Basic and acid dyes are also used largely for paper staining; perhaps most success is attained by adding a basic dye to the pulp first and then an acid dye when the dyes mutually fix each other. Direct cotton dyes are in use on better class paper and to some extent sulphide dyes, but the application of the latter is not easy.

The dyeing of wood, vegetable ivory, etc. follows the dyeing of wood chip in principle. It may however be done under pressure to obtain better penetration.

The fat or oil colours, which are soluble in oils, benzoline, alcohol, etc., are made from basic dyes by conversion into oleates, stearates or resinates; for example dissolve in 100 parts of hot water 1·25 parts of **Methyl Violet** and 5 parts of good soap. When cold add 2·5 parts of hydrochloric acid. The precipitated fat colour is collected, melted and the solid fat layer removed. Resinates may be prepared in a similar manner. Solutions of these fat colours in benzine are used in dry dyeing, and most of the colour manufacturers put out fat colours ready made. Oil colours are also used in boot polishes. For the colouring of fats and oils **Nigrosines** and **Sudan** colours are largely used. Dyes of this type are usually soluble in alcohol or methylated spirit, and along with many water-soluble basic dyes, which happen to be spirit-soluble also, are used for colouring varnishes and lacquers. In ink making basic and acid dyes are largely used, e.g., **Methyl Violet, Eosine,** etc. Gall inks are often shaded with acid dyes. Both acid and basic dyes find employment also in the colouring of soap, the dye being added to the kettle if it will stand boiling with alkali, if not it is added in subsequent mixing.

The production of **pigments** is an important branch of the colour manufacturing industry.

Acid and basic dyes are very largely used for production of **lakes** for wall-paper printing, coloured paper, textile printing (e.g., with albumen as fixing agent), for water colours, crayons, sealing wax, etc.

The lakes are produced by precipitating dyestuffs on to suitable bases, for example, barytes or heavy spar, aluminium hydrate, Blanc Fixe (precipitated barium sulphate) and for basic dyes white china clay or green earth. As precipitating agents barium chloride for all acid and most direct cotton dyestuffs, lead salts for **Eosines**, tannin and resin soaps for basic dyestuffs.

The choice of base or carrier is often more important than that of the dyestuff. For lakes of good covering power to be used in painting with oil or spirit varnish, lead sulphate, zinc white, Lithopone (a mixture of barium sulphate and zinc sulphide), barytes and red lead are used. Green earth is mostly used for lime colours for colour-washing walls. The finest and most transparent pigments such as are required for artist's colours and printing inks are made with alumina alone or with Blanc Fixe. **Example.** 20 parts aluminium sulphate ($2\frac{1}{2}$ °/₀ solution), 10 parts soda ash (Na_2CO_3, 10 °/₀ solution), 5—10 parts acid or direct cotton dyestuff (2 °/₀ solution), 10—100 parts barytes, 30—50 parts barium chloride (10 °/₀ solution). For a finer lake, barytes would be omitted or replaced by Blanc Fixe in small amount.

A few natural colouring matters also serve for lake production. The chrome and iron compounds with logwood give deep blue-black and brownish-black lakes. Brownish lakes are similarly obtained from **quercitron** and **fustic**, and red and maroon lakes from the **redwoods**, using alum, tin, chrome, etc.

The use of **dyestuffs for colouring foodstuffs, wines,** etc. is extensive and some care is demanded in their selection as many dyes possess more or less pronounced toxic properties. In general the following types are unsuitable: strongly basic dyes, nitro dyestuffs, dyes marketed as metallic salts of a harmful nature, e.g., zinc salts, or as double compounds with poisonous bodies, e.g., double oxalates like **Malachite Green,** dyes contaminated or liable to contain traces of poisonous bodies, e.g., arsenic used in **Magenta** manufacture which may be left in traces in the finished product.

The following well-known artificial dyestuffs are allowed by the United States Government for colouring foodstuffs: **Naphthol Yellow S, Orange I, Amaranth, Ponceau 3R, Erythrosine, Light Green SF yellowish, Indigo Disulphonate.**

Other well-known dyestuffs suitable are **Acid Magenta** (R.H.), **Auramine, Benzopurpurine, Metanil Yellow, Titan Scarlet** (R.H.).

Natural dyestuffs are usually preferred for colouring foodstuffs wherever applicable, and **cochineal, turmeric, annatto,** and **Persian berries** are largely used. **Tincture of Cochineal** is an aqueous-alcoholic extract of cochineal. The dried petals of **safflower** are used as a cosmetic.

Additional information may be acquired from the handbooks on dyeing in the Cambridge Technical Series. Reference may also be made to Jennison's *Lake Pigments from Artificial Colours*, Bersch's *Mineral and Lake Pigments*. Also the sections of Thorpe's *Dictionary of Applied Chemistry* (1912) on Dyeing, Lakes, Inks etc., and papers in the *Journal of the Society of Dyers and Colourists*. The literature issued by the various dyestuff manufacturers is replete with information on every branch of dyeing.

Note on Prices.—The prices at which dyestuffs are sold are variable and precise information is often difficult to acquire. The following short list of approximate prices in 1914 before the war will be of service to students possessing no inside knowledge of the trade:—Aniline 5*d.*—5½*d.* lb., carbolic acid (cryst.) 39°—40° C. 3½*d.*—5*d.* lb., picric acid (cryst.) 11*d.* lb., p-nitraniline 8*d.* lb., beta-naphthol 5⅝*d.* lb., phenylene diamine 1*s.* lb., resorcin 1*s.* 7*d.* lb.

Basic Dyestuffs. Auramine 1*s.*—2*s.* lb., Bismarck Brown 10*d.* lb., Chrysoidine 10*d.* lb., Malachite Green 1*s.* 6*d.*—1*s.* 9*d.* lb., Magenta 2*s.*—2*s.* 3*d.* lb., Methylene Blue 2*s.*—2*s.* 6*d.* lb., Rhodamine 1*s.* 6*d.*—2*s.* lb.

Acid Dyestuffs. Acid Violets, Acid Blues and Blacks 1*s.*—1*s.* 6*d.* lb., Oranges 5*d.*—1*s.* lb., Orange II 5*d.* lb., Ponceau 1*s.* lb., Scarlets 5*d.* (and upwards) lb., Crocein Scarlet 1*s.* 6*d.*—2*s.* lb., Victoria Scarlet 2*s.* 6*d.*—3*s.* lb., Tartrazine Pure 9*d.*—1*s.* 3*d.* lb., Naphthol Yellows 6¾*d.* lb., Fast Red A 7*d.* lb., Fast Red Extra 1*s.* 4*d.*—1*s.* 9*d.* lb., Soluble Blues 10*d.*—1*s.* 1*d.* lb., Brilliant Milling Green type 1*s.* lb., Patent Blue type 1*s.* 6*d.* lb., Indigo Extract 3*d.*—5*d.*, Refined 4*d.*—10*d.* lb.

Direct Cotton Dyestuffs. Benzopurpurines 6*d.*—9*d.* lb., Chrysophenine 9*d.* (and upwards) lb., Diamine Black RO, BH and developing blacks 1*s.* 6*d.*—1*s.* 10*d.* lb., Direct Blacks 1*s.*—1*s.* 6*d.* lb., Diamine Sky Blue and similar dyestuffs 11*d.* lb., Direct Pinks, Oranges, Greens, Reds and Violets 1*s.*—2*s.* 6*d.* lb., Yellows and Browns 7*d.*—1*s.* 2*d.* lb.

Mordant Dyestuffs. Alizarin 4½*d.* lb. (20 °/₀ paste), Alizarin Red (20 °/₀) 5½*d.* lb., Bluish Shade 6*d.*—7*d.* lb., Chrome Blues 1*s.* 4*d.*—1*s.* 6*d.* lb., Gallein (20 °/₀) 6½*d.*—7*d.* lb., Diamond Black type 7½*d.* lb., Acid Alizarin Blues 6*s.*—9*s.* lb.

Sulphur Dyes. Blacks 6*d.*—11*d.* lb., Browns 6*d.*—1*s.* lb., Blues 1*s.*—1*s.* 3*d.* lb., Greens 1*s.*—1*s.* 6*d.* lb., Yellows 6*d.*—8*d.* lb.

Vat Dyes. Indigo (20 °/₀ paste) 8*d.* lb.

Natural Dyestuffs. Indigo, Bengal 2*s.* 10*d.*—5*s.*, Oude 2*s.* 6*d.*—4*s.*, Kurpah 1*s.* 9*d.*—3*s.* 4*d.* lb., Logwood £6—£7¾ ton, Logwood Extract and Haematein Paste 20*s.*—30*s.* cwt., Haematein Crystals 50*s.*—65*s.* cwt., Barwood £7¼ ton, Camwood £16 ton, Peachwood £10 ton, Cochineal 2*s.*—2*s.* 6*d.* lb., Cudbear 4*d.*—5*d.* lb., Orchil Extract 25*s.*—30*s.* cwt., Persian Berries Extract 36*s.* cwt., Flavin 1*s.* 6*d.*—2*s.* 9*d.* lb., Fustic £5½ ton, Turmeric (Bengal) £21—(Madras) £26 ton, Annatto 6*d.*—1*s.* 6*d.* lb., Cutch and Gambier 20*s.*—36*s.* cwt.

CHAPTER XIV

COLOUR AND CONSTITUTION

WHEN a beam of white light falls upon a coloured surface part of the light may be reflected and part may pass through the material, a lowering in the intensity of light accompanying it. This will vary with different parts of the spectrum, consequently the reflected and transmitted light will have a different composition to the original source, it will therefore appear coloured. A portion of the light being absorbed in the coloured material, the corresponding colour or hue of a surface or substance is dependent on (1) the absorptive or selective power exhibited towards light, (2) the nature and composition of the light in which it is viewed, (3) the sensitiveness of the eye.

It thus happens that out of the indefinite number of light vibrations of different wave length, only the very limited range from 8100 A.U. (red) to 3600 A.U. (violet) (A.U.=Ångström unit $=10^{-8}$ cm.) is appreciated by the normal eye as colour sensations, and in some of the earlier civilisations the sense of colour perception was probably even still more limited. These facts cannot be ignored in seeking a relation between colour and chemical constitution.

The light absorbing properties of a substance are a principal factor in determining colour, and these absorption properties may well be related to the chemical constitution or arrangement of atoms in the substance. It is, however, only by investing the word "coloured" with a wider meaning than physiological limitations have caused to be attached to it, that a true theory of colour and constitution can be constructed. The practical usefulness of the chromophore theory, and its extension, the quinonoid theory of colour, has gained for them wide acceptance while still very incomplete and imperfect.

The portion of characteristic bands in the spectrum of a dye solution is readily ascertainable by spectroscopic methods. On addition of certain reagents modifications are caused which render further data available for a system of analysis of dyes known as Formanek's method, which has been applied also to vat dyes by Grandmougin.

Some substances only allow one colour of the spectrum to be transmitted, others two colours, or the spectrum may contain one or more dark bands, or a line spectrum along with dark absorption bands. Spectroscopic analysis of dyes depends on the differences shown in this respect. In nearly all cases the spectrum depends not only upon the nature of the dissolved substance, but also upon the thickness of the liquid, the concentration of the solution, the nature of the solvent and the temperature.

With many aromatic compounds absorption also takes place in the ultra violet portion of the spectrum, viz., among the invisible rays of light beyond the violet end of the spectrum, and absorption also may occur among the invisible light rays of the huge infra-red field adjoining the red of the visible spectrum. Under powerful electro-magnetic influence spectrum lines can be resolved into doublets, triplets, etc., and this phenomenon is known as the Zeeman effect. In the earlier attempts to relate colour to constitution absorption outside the range of visible light vibrations was neglected.

The physiological conception of white is obtained by

mixing the colours of the spectrum or by mixing two complementary colours of the spectrum. Since the combination of the colours of the spectrum gives white, then by absorption of one colour from the spectrum the complementary colour will be obtained.

The thickness of the coloured solution may not only cause a change in the intensity of the colour, but also a distinct change in the observed colour (dichroism). For example chrome alum solution in a thin layer is blue-grey and in a thicker layer violet-red. **Brilliant Acid Green 6B** is green in thin layers which gradually changes with increasing thickness through blue-green, blue, dark-blue, violet, purple, to red.

The nature of the solvent also plays an important part in the production of colour effects, since different solvents will have a different absorption value.

The temperature is also of importance, because increase in temperature tends to alter the absorption towards the red end, i.e., longer wave length, of the spectrum. A strong solution of copper chloride is blue, cold, but becomes green on heating.

Marked differences may also be observed between the colour of substances in the solid and liquid condition or between a solid and its solutions, e.g., crystals of p-nitrophenol or triphenylmethane are colourless, but when melted are yellow.

From a chemical standpoint the constitution of coloured substances is of great importance in connection with the theory of colour. By studying the constitution of a large number of organic compounds Graebe, Liebermann, Kostanecki, and Witt developed the **Chromophore Theory**, known as **Witt's Theory**. This theory states that the manner in which the atoms are grouped is the cause of compounds being coloured. The atom groupings are termed **Chromophores**. They may be divided into the following classes:

1. Ethylene group $>C=C<$ (indefinite position and number).
2. Ketone group $>C=O$.
3. Double bonded carbon and nitrogen

$$>C=NH \quad \text{or} \quad -CH=N.$$

4. Azo group $-N=N-$.
5. Nitroso group $-N=O$.
6. Nitro group $-N\begin{smallmatrix}\nearrow O \\ \searrow O\end{smallmatrix}$.
7. Thio group $>C=S$ or polysulphides.

Speaking generally, an increase in the number of chromophore groups has the effect of intensifying the colour.

The position of the chromophore group in the molecule is also of importance.

The following examples will illustrate the chromophore theory:

Fulvene $\begin{matrix} CH=CH \diagdown \\ | \qquad\quad C=CH_2 \\ CH=CH \diagup \end{matrix}$ is isomeric with benzene, the latter is colourless whilst fulvene is orange yellow.

In many cases it is necessary to have some ring system in the molecule in order to produce coloured bodies. Tetramethylbutadiene dicarboxylic acid

$$\begin{matrix} (CH_3)_2C=C-COOH \\ | \\ (CH_3)_2C=C-COOH \end{matrix}$$

is colourless, the tetraphenyl derivative

$$\begin{array}{l} (C_6H_5)_2—C=C—COOH \\ \qquad\qquad\quad | \\ (C_6H_5)_2—C=C—COOH \end{array}$$

is orange.

With hydrocarbons the following are examples of the ethylene chromophore structure:

$C=CH.CH_3$ Biphenylene-propylene (yellow)

$C=C$ Bibisphenvlene ethane (red)

$C=C$ Dinaphthylenediphenylene ethylene (violet red)

The ketone group as chromophore generally requires two or more keto groups or a certain ring system in order to produce colour.

$CH_3COCOCH_3$ yellow,

$—CO—$ colourless, $—C—$ ($\|$ O) orange.

The azo group as chromophore varies extremely in the intensity of colour which it produces in compounds.

Diazomethane $CH_2\Big\langle\begin{matrix}N\\ \| \\ N\end{matrix}$ is yellow.

Azobenzol $C_6H_5—N = N—C_6H_5$ red.

The nitroso group has very strong chromophore properties. Nitroso tertiary butane $(CH_3)_3C.NO$ in solution or in liquid condition is deep blue; under similar conditions nitrosobenzol C_6H_5NO is green.

The nitro group has a feeble chromophore character, this may be observed in the colour of nitrobenzene or nitronaphthalene. The position of the nitro group is not without influence. In solution p-nitrotriphenylfulgide is deep red, while the ortho and meta compounds are only feebly coloured.

The thiocarbonyl group, $= C = S$, has stronger chromophore properties than the carbonyl group $= C = O$.

Benzophenone is colourless, thiobenzophenone is a blue oil.

A considerable intensification of colour, due to the carbonyl group, is observable when other double bonds are present in the same compound. Phorone

$$(CH_3)_2C = CH . CO . CH = C(CH_3)_2$$

is yellow, the grouping of the atoms

$$>C = CH—CO—CH = C<$$

accounting for the colour.

The constitution of para- and ortho-benzoquinone or of other corresponding ortho- and para-quinones shows the effect of the proximity of double bonds in colour production.

O=C₆H₄=O (structure)

p-quinonoid

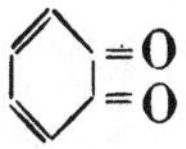

o-quinonoid

The para compound is yellow, the ortho red. The nearness of the two keto and also the two HC = CH groups causes the more intense colour in the ortho compound.

With quinones it often happens that two forms exist. By careful oxidation of catechol a colourless form may be produced which transforms into the coloured form, due to the following changes in structure:

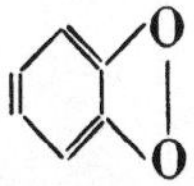

benzenoid (colourless) 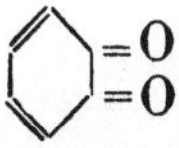 quinonoid (coloured)

There are many examples in which stereoisomerism causes a difference in the colour of compounds, corresponding to cis and trans forms.

```
R—C—H        R—C—H
  ||           ||
R′—C—H       H—C—R′
cis form     trans form
```

Diethoxy-naphthostilbene exists in two forms, the unstable, higher melting-point form is colourless, the lower melting form is yellow.

β-Brom allocinnamic acid is colourless, β-Brom cinnamic acid is yellow.

A similar difference may be noticed with diazo compounds.

$$\begin{array}{c} C_6H_5N \\ \| \\ KSO_3 . N \end{array} \qquad \begin{array}{c} C_6H_5N \\ \| \\ N . SO_3K \end{array}$$

Syn-Benzoldiazo sulphonate, orange — Anti-Benzoldiazo sulphonate, yellow

The origin of colour in a compound is due to the chemical nature of the chromophore groups. The coloured bodies that contain only chromophore groups, that is, coloured unsaturated compounds, these being termed **Chromogenes**, pass into colourless bodies by reduction, forming leuco compounds.

The following are examples:

Chromogene (coloured)	*Leuco compounds* (colourless)
$C_6H_5—N=N—C_6H_5$ Azobenzol (red)	$C_6H_5NH—NHC_6H_5$ Hydrazolbenzol
$O=C_6H_4=O$ Benzoquinone (yellow)	$HO . C_6H_4 . OH$ Hydroquinone
$O=C_6H_4—C_6H_4=O$ Diphenoquinone (yellow)	$HO . C_6H_4—C_6H_4 . OH$ Diphenol
$\begin{array}{l} C_6H_4\diagdown \\ \mid \qquad C=O \\ C_6H_4\diagup \end{array}$ Fluorenone (yellow)	$\begin{array}{l} C_6H_4\diagdown \\ \mid \qquad CCl_2 \\ C_6H_4\diagup \end{array}$ 9-9-Dichlorofluorene

By the introduction of new groups into chromogenes the colour may be decreased or intensified. Most chromogenes are relatively reactive bodies which may be more or less easily changed by chemical action. In some cases the introduction of a new group is accompanied by intramolecular change, such as the formation of quinone-like chromophores. The most important groups that affect the colour of a chromogene are the hydroxy, amino and substituted amino groups. These are known as **Auxochrome** groups. The auxochrome groups are unequal in value, e.g., the amino group develops a far more powerful action than the hydroxy group in para-nitraniline which is deep yellow, while p-nitrophenol is practically colourless. On the other hand **Benzo-purpurine 10B** (dianisidine coupled with 2 mols. α-naphthylamine-4-sulphonate) is carmine red, while **Benzoazurine G** of similar constitution, but a naphthol derivative, is blue. The auxochrome character of the hydroxyl group becomes more apparent by salt formation with bases. The replacement of hydrogen by alkyl or aryl groups also influences auxochrome groups. Indophenol $H_2N—C_6H_4—N = C_6H_4 = O$ is violet, whereas the dimethyl derivative is blue. Salt formation with chromogenes plays a very important part; if salt formation takes place in the auxochrome the colour becomes lighter, by salt formation in the chromophore the compound becomes darker. Ortho-aminobenzophenone is yellow, its hydrochloride is practically colourless.

Acridine is white, its salts are, however, yellow, also phenazine is yellow but it produces red salts. These are examples of salt formation in the chromophore.

Many coloured amines are tautomeric in such a way that the free base and its salts belong to different structural types. Generally the base contains the true benzol ring and the salts the quinonoid structure. In these cases salt formation is accompanied with intensification of colour. Para-amidoazobenzol is yellow, it gives bluish violet salts that dissolve in water with a red colour, para-para-diamidoazobenzol is similar; with the meta derivative salt formation reduces the colour intensity, as is also the case in the substances in which the auxochrome alone forms salts; m-m-diaminoazobenzol is dark orange, its salts are gold orange. In the case of the meta compound quinonoid structure cannot be set up. With the para compound transformation may take place, and since the para-quinonoid structure is a strong chromophore, such change intensifies the colour.

The triphenylmethane dyestuffs are particularly interesting. Hantzsch has been able to show that the free base and salt are of different types in this class of dyestuff. The free colourless bases, which he terms the leucohydrates, are to be considered as triphenylcarbinols, the coloured salts as quinonoid compounds.

The salts of **Crystal Violet** dissolve in excess of hydrochloric acid giving colourless solutions and are therefore derivatives of the normal carbinol.

Similar transformation has been observed with acridine derivatives. The colourless methyl-phenyl-acridol

CH_3

N

C

C_6H_5 OH

gives a dark brownish-black iodide having the following constitution:

CH_3 I

N

C

C_6H_5

With most coloured phenols the auxochrome nature of the hydroxyl group is made obvious only by salt formation. The colour increasing on addition of alkali. It is possible that this change is due to transformation to the quinonoid structure

$$C_6H_4\begin{matrix}-OH\\-NO_2\end{matrix} \rightarrow C_6H_4\begin{matrix}=O\\=NO_2Na\end{matrix}$$

This is however hardly probable since meta-nitrophenol gives the same colour change without the possibility of quinonoid structure.

By salt formation of amines with acids without change in the chromophore, the colour is decreased.

By salt formation of hydroxyl (phenolic) groups without change in the chromophore, the colour is intensified.

In addition to the colour change by salt formation there is also the change due to **Halochrome** groups (Baeyer). The halochrome groups of most importance are oxygen derivatives, in which the oxygen apparently possesses basic properties to a mild degree, passing from the divalent to the tetravalent basic form. Many ketones having the group

$$-CH=CH-CO-CH=CH-$$

with dry hydrochloric acid gas give a darker coloured compound. It has been shown that in these compounds the double bond is not attacked. The salts are easily decomposed, either by addition of water or placing in a vacuum.

Thianthrene $C_6H_4\langle\begin{matrix}S\\S\end{matrix}\rangle C_6H_4$ has strong halochrome properties, dissolving in concentrated sulphuric acid and thereby giving a fine blue solution.

Groups or atoms that cause an intensification of colour when introduced into compounds have been called **Bathochromes**, and those that decrease the colour of a compound **Hypsochromes**.

Alkyl halides when combined with an auxochrome decrease colour, e.g., m-nitro-dimethyl aniline is yellow, but its methyl bromide addition product is white. If the alkyl halides are attached to the chromophore, then as a rule the colour is intensified, e.g., acridine is colourless and its methyl iodide addition product is red.

Acylation (e.g., Acetylation and Benzoylation) always gives hypsochrome properties, whether acting upon amines or hydroxy groups. In homologous series an increase in molecular weight is accompanied by increased intensity of colour, and the introduction of alkyl groups into ring compounds shows that these groups possess a bathochrome character. The naphthalene derivatives are more intensely coloured than the corresponding benzene derivatives. Halogenation is accompanied as a rule with bathochrome functions. The intensification is not proportional to the number of halogen groups introduced, but the position of the halogen groups is important. The sulphonic acid group to a mild degree acts both as batho- and as hypso-chrome.

From what has been stated already, it will be understood that what is called the quinonoid theory of colour, while it does not deny much that is assumed by the chromophore theory, definitely attempts to relate the structure of dyes to either ortho- or para-quinones. It is assumed that colour change, e.g., as the effect of a solvent or by salt formation, is accompanied by a change in the primary structure of its molecules, the setting up of a quinonoid type in place of benzenoid structure being accountable for a change from colourless to coloured, and *vice versa*. A large amount of experimental evidence collected by Hantzsch and others may be claimed in support of this view. It is however a theory for aromatic compounds only; it ignores the immense range of light vibrations in the infra red, thus being adopted on a limited and arbitrary definition of colour, but much may be said for its usefulness as a working hypothesis. The identity of mononitrosophenol and monoquinone oxime, and also a similar identity in corresponding naphthalene derivatives, has led to the assumption of isomeric changes fitting well with the quinonoid theory.

OH (ring) NO $\rightleftarrows$ O (ring) NOH also OH (naphthalene ring) $N.NC_6H_5$ $\rightleftarrows$ O (naphthalene ring) $N.NH.C_6H_5$

Investigations into the isomeric coloured and colourless ethers of **Fluorescein** and **Phenolphthalein** have also resulted in strong evidence of such isomeric change from benzenoid to quinonoid type, according to whether coloured or colourless. The quinonoid theory has acted as a check on the over-rapid application to dyestuffs of simple ionic explanations of colour phenomena, which often ignore the actual chemical processes involved, e.g., in the case of Fluorescein and Phenolphthalein. The student however must not build too much upon the rigidity of chemical conceptions, and hypotheses framed to overcome some difficulty or inconsistency in formulae. Thus the usual formulae of structural chemistry having proved unequal to the strain of explaining the identity of p-nitrosophenol and p-quinone oxime, use was made of the conception of isomeric change; this affords an excellent working hypothesis, but the reality of any such change must still be doubtful in the extreme. The quinonoid theory as at present applied is open to many grave fundamental objections, more so than such a theory of colour and constitution as is put forward by Baly, who claims that the **molecular force field theory** gives a rational explanation of the phenomena of light absorption and attendant colour. As a working hypothesis for use in synthetic dyestuff chemistry however, the latter theory cannot claim to be anything like so fruitful as the quinonoid theory. To a student of applied chemistry such distinctions are valuable, and must be borne in mind. The force field theory has been developed from the notion of chemical affinity being the external evidence of the force fields surrounding atoms or molecules. It follows from what is known as the Zeemann effect, viz., the resolution of spectrum lines into doublets, triplets, etc., in a powerful magnetic field, that each atom must form the centre of an electro-magnetic field due to the rotation of its component electrons. If two molecules, the force fields of which are of opposite type, are brought together, the two force fields will condense with escape of energy. When a force field is closed the body cannot react. If the free force lines of another body are caused to interpenetrate the closed fields of the former, they will become opened up. In the case of the closed force field of a base, an acid is the most suitable substance to open it up. The fact that a salt is produced is regarded as of no consequence, e.g., in the case of phenol and alkali. Phenol in alcoholic solution shows two absorption bands in the ultra violet, and in the alkaline solution it also exhibits two absorption bands. The frequency difference between these four bands is about 160. A strong absorption band in the infra red also has a frequency of 160. Similarly with the nitrophenols, in neutral and alkaline solutions they exhibit different bands, but in each case the different absorption bands in the two solvents are directly connected by the frequency relation, and the nitrophenols must therefore possess exactly analogous constitutions in neutral and alkaline solutions. Such results are quite at variance with the assumption of isomeric quinonoid change occurring in these cases.

Additional information may be obtained from the following sources:

Baly, "Colour and Constitution," *Jour. Soc. Dyers*, p. 39, 1915.

Green, "Quinonoid Addition as the Mechanism of Dyestuff Formation," *Jour. Chem. Soc.*, p. 925, 1913.

Bearder, "Fluorescence," *Jour. Soc. Dyers*, p. 270, 1911.

May, "Origin of Colour in Organic Compounds," *Chem. News*, pp. 283 and 295, 1907.

von Baeyer, "On Aniline Colours," *Zeit. f. ang. Chem.* 1906, abstract in *Jour. Soc. Dyers*, p. 335, 1906.

Kauffmann, "On the Relation between Constitution and Colour," *Ahren's Sammlung* 1907, F. Enke, Stuttgart.

Formanek, "Spektralanalytischer Nachweis kunstl. organ. Farbstoffe," Springer, Berlin.

CHAPTER XV

NITROSO AND NITRO DYESTUFFS

Nitroso Dyestuffs.—A small group obtained by the action of nitrous acid ($NaNO_2 + HCl$) on phenols. The introduction of the chromophore group NO gives colour, and when ortho- to the hydroxy group also gives mordant dyeing properties. These dyes are usually applied in conjunction with iron, giving dull green shades of good fastness to light and washing, being much used in printing of fabrics.

Fast Green O (paste) (M.).

Dark Green (paste) (C.), **Resorcine Green**, etc.

OH, NO, ON, OH (benzene ring) or O, =NOH, HON, O (quinone ring)

Gambine R (R.H.), 1-oxy-2-nitrosonaphthalene.

Gambine Y (R.H.), 2-oxy-1-nitrosonaphthalene.

Gambine B (R.H.), 1-nitroso-2.7-dioxynaphthalene.

Naphthol Green B (C.) is obtained by use of a phenol sulphonic acid from which NaO_3S–$C_{10}H_5$(NO)(OH) (naphthalene ring) is obtained by the usual method. It is marketed as an iron salt, having one atom of iron (Fe″) united with two of the above molecules and is used mainly for wool.

Nitro Dyestuffs.—A few acid products coloured yellow in virtue of possessing the chromophore group NO_2 are used as dyes.

Picric Acid or symmetrical tri-nitro-phenol, described previously among the intermediate products, finds a limited use as an acid dye.

Martius Yellow, Naphthalene Yellow, etc., is prepared similarly to picric acid, i.e., from the 2.4-disulphonic acid of α-naphthol.

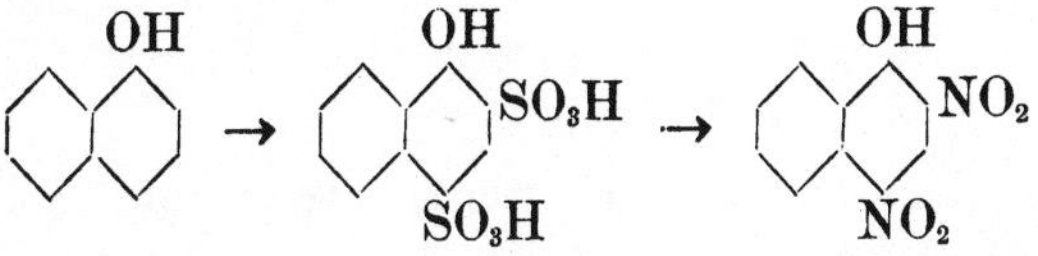

OH → OH, SO_3H, SO_3H → OH, NO_2, NO_2

It appears on the market as ammonium, sodium or calcium salt.

Naphthol Yellow S.—Obtained by nitration of 1-naphthol-2.4.7-trisulphonic acid or of the 1.2.7- or 1.4.7-disulphonic acids. It is obtained as potassium or sodium salt.

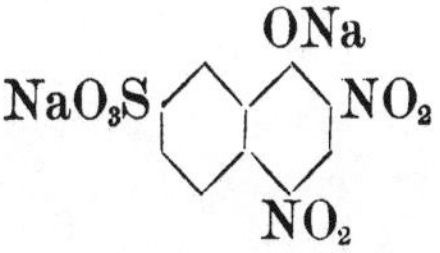

ONa, NaO_3S, NO_2, NO_2

It is very extensively used as an acid dye, and has almost entirely replaced **Martius Yellow.**

The standard of fastness in the nitro dyes is only fair.

The nitro dyestuffs, like the nitroso dyestuffs, are destroyed on reduction, forming colourless amido compounds.

CHAPTER XVI

AZO DYESTUFFS

Azo dyes contain the chromophore group —N = N—. (The azo group in the formulae under this section is expressed by the common contraction $—N_2—$.) This large class of dyes has been obtained by coupling diazo compounds with phenols or amines or suitable derivatives. Coupling depends on the removal of an atom of hydrogen from the ring by formation of free acid with the acid radicle of a diazo salt, whereby the diazo body joins on to the ring, giving quantitative substitution by formation of an azo compound. E.g., diazobenzene chloride and β-naphthol give **Sudan I** thus:

$$C_6H_5 . N_2 . Cl + C_{10}H_7OH = HCl + C_{10}H_6(N_2 . C_6H_5)OH$$

To obtain ready coupling, the hydrochloric acid must be removed, and this is done by coupling in presence of caustic soda, sodium carbonate or sodium acetate. Usually the diazo compound is in solution, and it is desirable that the other component, phenol or amine, should also be in solution. It is usual to dissolve phenolic compounds with the aid of a theoretical amount of caustic soda, and then add sodium carbonate in amount sufficient to unite with the whole of the hydrochloric acid set free in coupling. On account of the liability, especially the case with simpler diazo bodies, to form anti-diazo compounds which couple with great difficulty, an excess of strong alkali is to be avoided.

Coupling with amines is generally done in acid solution, as acid is usually required to dissolve these bodies. Such coupling is usually more difficult than with phenolic compounds, and it is often the case that two couplings are necessary to obtain a certain disazo dye, i.e., one containing two azo groups, one coupling being in a phenolic substituted ring, and the other in an amino substituted ring. The latter being more difficult must be done first, as increased complexity or molecular weight, such as would be caused by coupling first on the phenolic side, renders it all the more difficult. In certain cases, e.g., the phenylene and tolylene diamines, where the amine is easily soluble in water alone, coupling may be done in neutral or even alkaline solution, the latter being commonly used in works practice. In other cases non-aqueous solvents are used, e.g., in coupling m-sulphanilic acid on to diphenylamine, the latter is dissolved in alcohol, giving **Metanil Yellow**. Sodium acetate is often added during coupling of amines to remove the free hydrochloric acid that is used to dissolve these components, the free hydrochloric acid in solution being replaced by acetic acid, and at the same time sodium chloride is formed. Commercial dyestuffs are produced containing not only one or two azo groups but also trisazo dyes containing three, and tetrakisazo dyes containing four azo groups. Thus the following series of couplings may take place with aniline as the first component (the body diazotised) and resorcin as second component:

OH, OH (resorcin ring)	OH, OH, $N_2 . C_6H_5$ monazo **Sudan G**
C_6H_5 N_2, OH, OH, $N_2 . C_6H_5$ disazo	$C_6H_5 . N_2$, OH, $N_2 . C_6H_5$, OH, $N_2 . C_6H_5$ trisazo

With each increase in the number of azo groups the shade is deepened, and also coupling becomes more difficult. These are the main reasons why coupling is rarely carried on to give tetrakisazo bodies, and practically never to give pentazo dyes. Also a diminished affinity for fibres often becomes evident in tetrakisazo dyes, especially with direct cotton dyes.

Most influential in coupling is the character and position of groups already substituted in the ring to be coupled with a diazo body. Coupling cannot take place in an unsubstituted aromatic nucleus. Thus in the case of **Sudan I**, quoted above, coupling is possible only in the nucleus containing the hydroxy group. As has been already indicated, hydroxy or amido groups or substituted amido groups, e.g., $—N(CH_3)_2$, are required in the nucleus before coupling can take place. Where this primary condition is fulfilled, other groups, the presence of which alone will not induce coupling, may now distinctly favour it; notably electro-negative groups, e.g., halogens, nitro, methyl, etc., groups in amines are favourable to coupling. Thus aniline as second component couples with some difficulty, toluidines couple more readily, with less tendency to form diazoamido compounds, while xylidines couple, giving azo compounds quite easily, as do also naphthylamines. The stability of diazo compounds is also increased by the presence of electro-negative substituents; e.g., diazotised p-nitraniline is more stable than diazobenzene. The stability of these bodies is important as regards time and temperature in coupling. Diazobenzene must be coupled in ice-cooled solution, but diazotised p-nitraniline may be coupled in ordinary cold solution, and certain other diazo compounds are stable in presence of hot water, e.g., diazo compound from o-anisidine. Generally, coupling is done in ordinary cold solution, ice being added where necessary.

The position in which coupling takes place in the ring of the second component must now be discussed. A general rule is that substitution by diazo groups takes place in the para position to the hydroxy or amido group where possible, and if prevented by the presence of another substituent, the ortho position to the auxochrome group is taken. Ortho-coupling most commonly occurs with phenols in alkaline solution. The example of diazobenzene chloride and resorcin given above may be referred to here also in illustration. Special modifications of the above rule must now be stated.

In the class of benzene derivatives, ortho- and para- substituted di-amido- and di-oxy-benzenes do not couple with diazo bodies under ordinary conditions, and in practice only meta derivatives of this type are in use as second components, e.g., resorcin, m-phenylene diamine, m-tolylene diamine, etc. Derivatives of aniline with substituents in the para position, e.g., sulphanilic acid, also refuse to couple with diazo compounds.

In the naphthalene series other modifications prevail. Coupling with α-naphthol or α-naphthylamine occurs in the para position to the auxochrome. Where the para position is already substituted, or if already coupled, the ortho position is assumed. Also it is found that if the para position be actually free to couple of itself, the presence of a substituent group in either adjacent position will exert steric hindrance, and prevent coupling at this point. For example:

(i) α-naphthol behaves thus:

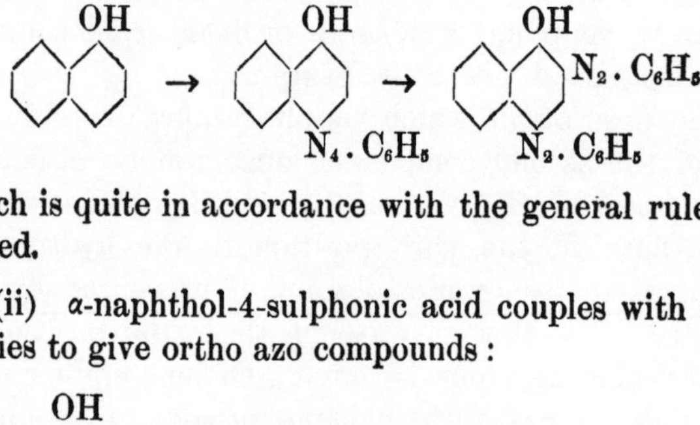

OH → OH → OH

$N_2 . C_6H_5$

$N_4 . C_6H_5$ $N_2 . C_6H_5$

which is quite in accordance with the general rule first stated.

(ii) α-naphthol-4-sulphonic acid couples with diazo bodies to give ortho azo compounds:

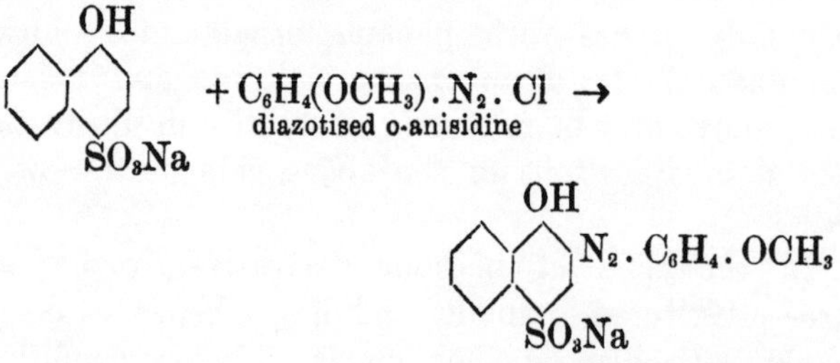

OH

$+ C_6H_4(OCH_3) . N_2 . Cl$ →
diazotised o-anisidine

SO_3Na

OH

$N_2 . C_6H_4 . OCH_3$

SO_3Na

Azoeosine G (By.)

(iii) If the 1 . 5-naphthol sulphonic acid be employed,

coupling can only take place in the ortho position to the oxy group:

$$\text{(OH, NaO}_3\text{S-naphthalene)} + C_6H_5 . N_2 . Cl \rightarrow \text{(OH, NaO}_3\text{S-naphthalene)}-N_2 . C_6H_5$$

Cochineal Scarlet G (Sch.)

Steric nindrance of the peri substituent, i.e., the sulphonic group, here prevents position 4 being filled.

(iv) Again where position 3 from the auxochrome is filled as in NaO_3S(OH-naphthalene)SO_3Na coupling takes place in the ortho position to the oxy group. Thus with diazotised m-xylidine **Palatine Scarlet** (B.) is obtained:

$$\text{NaO}_3\text{S-(OH-naphthalene)}\begin{matrix} -N_2 . C_6H_3 . (CH_3)_2 \\ SO_3Na \end{matrix}$$

In the case of β-oxy and amido naphthalene and their derivatives, coupling always takes place in the adjacent α position. E.g., aniline when diazotised couples with β-naphthol in the position shown by the formula for **Sudan I.** (See page 136.) Or again, if diazotised sulphanilic acid be coupled on to β-naphthol, **Orange II** is obtained:

$$\text{(naphthalene)} \begin{matrix} N_2 . C_6H_4 . SO_3Na \\ OH \end{matrix}$$

Coupling with di-amido- and oxy-naphthalenes obeys two rules: (i) 1.2 and 2.1 amido-naphthols cannot be coupled; (ii) where the two auxochrome groups are in different nuclei, coupling may take place in each nucleus under the conditions laid down above.

For example, **Palatine Black** (B.) is obtained from H-acid by coupling on diazotised sulphanilic acid, first in acid solution (being the more difficult coupling), followed by coupling diazotised α-naphthylamine to the monazo dye thus prepared, after making the solution alkaline.

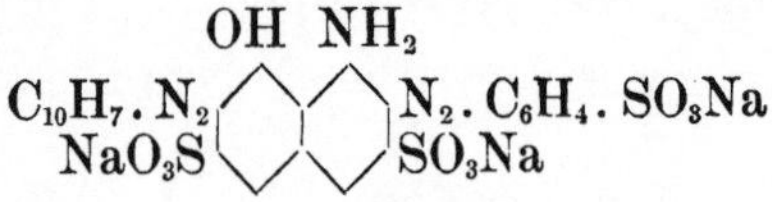

$$\begin{array}{c} \quad OH \quad NH_2 \\ C_{10}H_7 . N_2 \text{[naphthalene]} N_2 . C_6H_4 . SO_3Na \\ NaO_3S \qquad\qquad SO_3Na \end{array}$$

In certain cases where coupling is difficult, copper salts are added to assist the operation.

Beside the effect of the number of azo groups on the shade, the nature, position and number of auxochrome groups also influences colour.

Bluer or deeper shades are obtained from hydroxy groups than from amido groups, while methoxy and ethoxy groups tend to give brilliant shades, especially in certain positions, e.g., ortho to the chromophore group —N : N—.

Thus aniline, coupled with 1.4-naphthol sulphonic acid, gives an orange dye not in use now, while o-anisidine, in place of aniline, gives **Azoeosine G** (By.) (see p. 140), which dyes bright bluish red shades. The fastness of dyes to light is always important, and the use of chlorinated derivatives almost invariably increases resistance to fading, and frequently also to other agencies, e.g., oxidation, alkalies and acids. Thus dichlorobenzidine is used instead of benzidine, e.g., in certain **Dianol Reds** (Lev.), and chlorinated naphthol sulphonic acids are used as second components, with dianisidine as first, in the **Chlorazol Blues** (R.H.). Fastness to acids and alkalies is largely dependent upon the ability of such reagents to attack auxochrome groups. Thus the entrance of sodium into a hydroxy group usually modifies the colour and fastness. This occurs more easily where the auxochrome oxy group is para to the azo group, and ortho-oxy-azo dyes are much more valuable on account of offering greater resistance to alkalies. Where however coupling takes place in more than one stage, e.g., in making disazo and trisazo dyes, if it is desired to diazotise an amido azo compound for this purpose, such compounds, with the amido group ortho to the azo group, e.g., amido azo derivatives of β-naphthylamine, offer great resistance to diazotisation, while other amido groups may be readily diazotised. Hence to obtain successive coupling with amines and rediazotisation of the amido azo compounds formed, p-amido-azo compounds are desirable, as they allow of ready diazotisation, e.g., α-naphthylamine and its derivatives are frequently used on this account, as in the case of **Naphthylamine Black D** (C.). This dyestuff is obtained by coupling diazotised α-naphthylamine 3.6-disulphonic acid to α-naphthylamine, the resulting p-amido-azo compound being rediazotised and again coupled with α-naphthylamine.

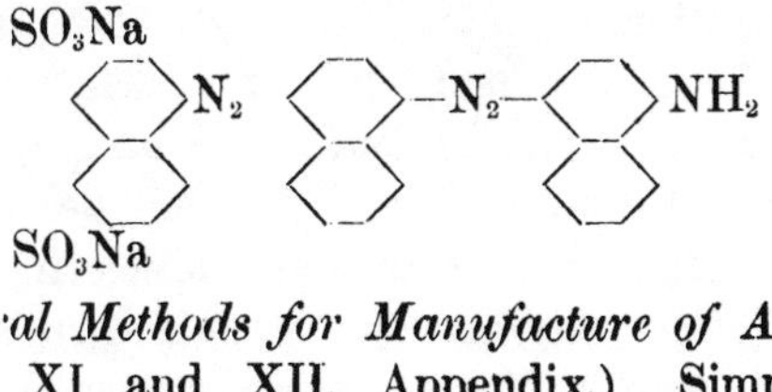

SO_3Na

N_2 — N_2 — NH_2

SO_3Na

General Methods for Manufacture of Azo Dyes.—(Figures XI and XII, Appendix.) Simple amines are dissolved as hydrochlorides in water, and diazotised by rapid addition of sodium nitrite in slight excess of the theoretical amount, to avoid possible formation of diazo amido bodies. The calculated quantity of hydrochloric acid, necessary to decompose the nitrite, is previously added to the amine solution. The reactions are done in wooden vats with rotating paddle stirrers (occasionally shovel stirrers). The large coupling vat is on the ground, and supported on a platform above are the diazotising vat and the vat containing the acid or alkaline solution of the second component to be coupled. A small nitrite vessel is placed above the diazotising vat, and pipe connections are fitted from vessel to vessel, leading down ultimately to the coupling vat. The nitrite pipe is carried to the bottom of the diazotising vat to avoid loss by nitrous fumes. The amido-sulphonic acids are often mixed with nitrite solution and diazotised by running hydrochloric acid into the mixture. The formation of dyestuff in the coupling vat varies in period; in the case of simpler bodies it often takes place at once, in other cases several days' stirring is required for complete coupling. For convenience in working, the contents of the coupling vat may be passed on to a tank or "monteju," from which it may be pumped or blown with compressed air to the filter press.

Usually, after coupling, azo dyes are largely in solution, and salt is added where necessary to salt out or precipitate the dye. Filtration is accomplished by means of a filter press, for a detailed description of which some manual of chemical engineering may be consulted. It consists, however, essentially of a number of plates of wood or metal, with distance frames and pieces of cloth between, the whole when screwed up forming a horizontal block of separate chambers, into which the liquor containing precipitate can be forced through tunnels running through the block. The precipitate is retained in the chambers, while clear filtrate passes out through the cloth layer on each side to an outlet tunnel. For acid liquors the press is constructed of wood. Iron is used for neutral and alkaline liquors. Filter cloth materials consist of stout-ribbed cotton for general use where neutral or alkaline liquors are in question; flannel for acids, and camel's hair cloth for strong acids. Certain dye manufacturers prefer small presses (4 feet length), others use much larger ones. After filtering, the press is unscrewed, the press cakes turned out of the cloths on to trays, and dried in vacuum ovens at 75° C. This later method gives better products than the old stove drying. So-called "concentrated brands" of azo dyes are obtained by the use of hydraulic pressure on the filter cakes, thus squeezing out a good deal of salt.

Finally the dyes are ground in a ball or roller mill, this being an iron cylinder with a screw lid, which is revolved so as to cause enclosed iron balls or rollers to pound the dry cakes of dye put inside to powder. Although so strictly a scientific process, the various batches of an azo dye are never quite the same strength, and a suitably varied amount of common salt or sodium sulphate is generally ground with the dye, in order to turn out a standard strength of product. It is common also to grind a little soda with dyes of an acid character, to facilitate solution. More serious adulteration may readily be practised in the grinding of dyes than is justified by the necessity of turning out a standard product

Occasionally metallic salts or mordanting substances may be ground in with dyes to give some special properties when dyed, e.g., ortho-oxy-azo dyes (see p. 154). The mixing of dyes to obtain some special shade or mixture, as for union dyeing, is also frequently done in the grinding.

General Properties of Azo Dyes.—Azo dyes may be of various dyeing classes, e.g., basic, acid, mordant, and direct cotton dyes, but in all cases on reduction are decolorised by splitting up at the azo groups, giving amido compounds thus:

$$R—N:N—R' + 4H = R—NH_2 + R'—NH_2.$$

Reduction methods of both qualitative and quantitative analysis are based on this reaction, which takes place usually in hot aqueous solution with one of the following reducing agents: zinc dust and acid or alkali, titanous or stannous chloride and hydrochloric acid, hydrosulphite compounds, etc. Such reduction methods are used for stripping azo dyes from dyed cloth, and for discharge styles in calico printing. The colours of azo dyes in strong sulphuric acid solution vary, and are often characteristic.

Insoluble Azo Dyes.—Certain of these, oxy-azo compounds, are made for use as spirit colours, e.g., the **Sudans.**

Brilliant Lake Red R (M.), obtained from aniline and β-oxynaphthoic acid, is an example of azo dyes used only for lake manufacture. Many water soluble dyes are obtained as lakes, however, by conversion into sparingly soluble salts, such as calcium and barium salts.

There is also a small class of insoluble azo dyes produced on the fibre, of which the chief is p-nitraniline red or **Para-Red.** Cotton is steeped or padded in sodium β-naphtholate solution, and, after squeezing out evenly, dried. It is now treated in a 1 per cent. solution of diazotised p-nitraniline, when coupling takes place, the insoluble azo dye formed on the fibre being firmly fixed there. It is much used for cheap imitation

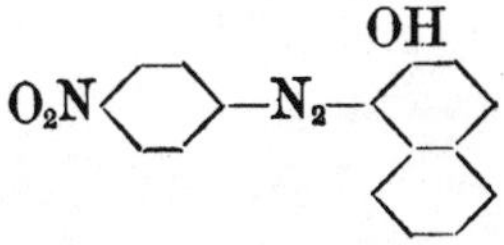

OH

O_2N—N_2—

Turkey Reds. β-naphthol is almost exclusively used as second component (**Naphthol AS** (G.E.), the anilide of β-oxynaphthoic acid, offers certain advantages over β-naphthol as a "grounding" or "prepare"), but a variety of diazo compounds are in use as well as p-nitraniline. The most important of these, and the respective colours of the dyes obtained when coupled with β-naphthol are as follows:

p-nitraniline → bright red.
m- or o-nitraniline → orange.
α-naphthylamine → claret.
benzidine → puce.
dianisidine → blue.
o-anisidine, chloranisidine,
or p-nitroanisidine → reds.

Azophor Red PN, Nitrazol C, Nitrosamine Red, Azogen Red and **Benzonitrol** are pastes containing diazotised p-nitraniline rendered stable, e.g., with strong acids.

Basic Azo Dyes.—Of the basic class only a few are in use. To obtain the azo bases in a water soluble form they are converted into hydrochlorides.

Aniline Yellow or p-amidoazobenzene

$$C_6H_5 . N_2 . C_6H_4 . NH_2,$$

and its homologue p-amido-azo-toluene are used for colouring fats (e.g., butter substitutes) and varnishes.

Chrysoidine is a well-known yellow basic dye obtained from diazobenzene chloride and m-phenylene diamine by coupling and crystallising the hydrochloride from solution in hot water.

$$C_6H_5 . N_2 . C_6H_3(NH_2) . NH_2 . HCl$$

Strongly basic dyes suitable for leather are obtained by use of strongly basic derivatives, e.g., (i) p-amido-benzyldimethylamine coupled with β-naphthol is the main constituent of **Tannin Orange R** (C.)

$$(CH_3)_2NH_2C . C_6H_4 . N_2 . C_{10}H_6 . OH$$

(ii) m-amido-phenyl-trimethylamine coupled with resorcin gives **Azophosphine GO** (M.)

$$N(CH_3)_3Cl . C_6H_4 . N_2 . C_6H_3(OH)_2$$

The **Janus** dyes (M.) are strongly basic dyes dyeing cotton direct. **Janus Red B** is obtained by diazotising m-amido-phenyl-trimethylamine, coupling with m-toluidine and again coupling the amido-azo compound thus formed with β-naphthol.

Bismarck Brown is the hydrochloride of the disazo

base obtained by treatment of metaphenylene diamine with nitrous acid in aqueous solution. The calculated amount of nitrite is used to get one-third of the diamine completely diazotised in both amido groups. Coupling takes place with the remaining diamine to give the dye, which mainly consists of the disazo compound shown.

N_2 — $C_6H_3(NH_2)_2$ (NH_2, NH_2)

C_6H_4

N_2 — $C_6H_3(NH_2)_2$ (NH_2, NH_2)

It is still largely used on animal fibres, vegetable fibres, and leather. Other brands are made from m-tolylene diamine. Other names for these products are **Vesuvine** (B.), **Manchester Brown, Leather Brown, Phenylene Brown,** etc.

By diazotisation of **Safranine** (see p. 231) in one amido group and coupling, strongly basic monoazo dyes are obtained. Among others are the following, the second component used along with diazotised Safranine being given in each case:

phenol—**Diazine Black** (K.).
β-naphthol—**Indoine Blue, Janus Blue** (M.), etc.
dimethylaniline—**Diazine Green** (K.), **Janus Green** (M.) (many other names).

Mordant Azo Dyes.—In this section only the simple mordant dyes incapable of application by other methods are considered. The acid mordant dyes, a much more numerous and important series of azo compounds, are dealt with later (see p. 153). The required grouping in the mordant dyes is generally obtained by use of salicylic acid as an end component. **Alizarin 2 G** (M.) is obtained by coupling with diazotised m-nitraniline.

$$\underset{NO_2}{C_6H_4} \cdot N_2 \cdot \underset{COOH}{C_6H_3} \cdot OH$$

It is also sold as **Alizarine Yellow in paste, Mordant Yellow** (B.), **Anthracene Yellow 2 G** (C.), etc.

Alizarin Yellow R (M.), also sold under many other titles, is obtained by use of p-nitraniline.

$$O_2N \cdot C_6H_4 \cdot N_2 \cdot \underset{COOH}{C_6H_3} \cdot OH$$

Other mordant azo dyes and the components used in their manufacture are:

Prage Alizarin Yellow G (Ki.) from meta-nitraniline and β-resorcylic acid.

Azogallein (G.) from p-amido dimethylaniline and pyrogallol. Deep violet on chrome mordant.

Azochromine (G.) from p-amidophenol and pyrogallol. Brown on chrome mordant.

Diamond Flavin G (By.), obtained by coupling tetrazotised benzidine with one molecule of salicylic acid and boiling to decompose the other diazo group.

$$\begin{array}{l} C_6H_4 \cdot OH \\ | \qquad\qquad\quad COOH \\ C_6H_4 \cdot N_2 \cdot C_6H_3 \cdot OH \end{array}$$ Yellow on chrome mordant.

Mordant Yellow GRO (B.) is obtained as for Diamond Flavin by treatment with bisulphite instead of merely boiling with water.

This last body is water soluble; the others are scarcely to be reckoned soluble, and are usually sold as pastes. They are dyed in a manner similar to **Alizarin** and other simple mordant dyes, and give shades of almost similar excellent fastness.

Anthracene Yellow C (C.), also of this class, is obtained from thioaniline and two molecules of salicylic acid.

Acid Azo Dyes.—The commercial dyes of this class are very numerous, including every shade, and frequently attaining to a considerable degree of fastness. They are considerably used in wool dyeing, and to a less extent in silk dyeing. The number and variety is so large that only a small selection of well-known products and their components can be given here.

Monazo Acid Dyestuffs.

Colour.

Orange. **Orange G**

aniline → 2-naphthol-6.8-disulphonic acid.

Orange red. **Ponceau G** (B., M., etc.)

Orange R (R. H.)

aniline → 2-naphthol-3.6-disulphonic acid.

Bright bluish scarlet. **Ponceau R, 2 R G,** etc. (Xylidine Scarlet.)

xylidine → 2-naphthol-3.6-disulphonic acid.
(m- or mixed xylidines according to brands.)

Bluish red. **Azo Acid Red B** (M.), **Lanafuchsin 6 B,** etc. (C.), **Sorbine Red 2 B, G** (B.), etc.

acetyl p-phenylene diamine → 1-naphthol-3.6-disulphonic acid.

Bordeaux. **Fast Red B or Bordeaux B,** etc.

α-naphthylamine → 2-naphthol-3.6-disulphonic acid.

Bright red. **Crystal Scarlet or Ponceau 6 R,** etc.

α-naphthylamine → 2-naphthol-6.8-disulphonic acid.

Colour.

Golden yellow. **Metanil Yellow** (various brands)
m-sulphanilic acid → diphenylamine.

Orange. **Orange IV.**
p-sulphanilic acid → diphenylamine.

Orange. **Orange I.**
p-sulphanilic acid → α-naphthol.

Bluish red. **Azofuchsin G** (By.),
p-sulphanilic acid → 1 . 8-dioxynaphthalene-4-sulphonic acid.

Brown. **Naphthylamine Brown** (B.)
naphthionic acid → α-naphthol.

Red. **Fast Red A**
naphthionic acid → β-naphthol.

Red. **Amaranth** (C.)
naphthionic acid → 2-naphthol-3 . 6-disulphonic acid.

Red. **Azorubine** (various brands)
naphthionic acid → α-naphthol-4-sulphonic acid.

Violet. **Lanacyl Violet B** (C.)
1-amido-8-naphthol-3 . 6-disulphonic acid (H acid)
→ ethyl-α-naphthylamine.

Blue. **Lanacyl Blue 2B** (C.)
1-amido-8-naphthol-3 . 6-disulphonic acid (H acid)
→ 1-amido-5-naphthol
(coupled in alkaline solution).

The student should write full formulae for each of the above dyestuffs, attending carefully to the rules for coupling laid down previously. The effect of different groups on shade may then be noted. **Metanil Yellow** is rather sensitive to acids, turning red. It is much used in the paper industry. Where detailed information as to fastness, levelling properties, etc., is available, from the makers' pattern cards for example, some comparison of these properties should also be made. See also **Azoeosine G**, **Cochineal Scarlet G**, **Palatine Scarlet**, **Orange II.**

Disazo Acid Dyestuffs.

The direction of the arrows indicates the order of coupling.

Colour.

Brown. **Fast Brown** (By.)

naphthionic acid → resorcin ← naphthionic acid.

Blue Black. **Naphthol Blue Black S** (C.) (many other brands)

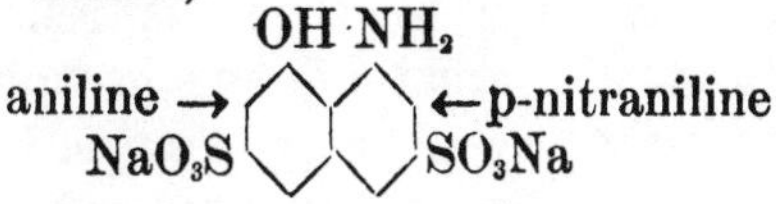

$$\text{aniline} \rightarrow \underset{NaO_3S\qquad SO_3Na}{\overset{OH\cdot NH_2}{\text{[naphthalene ring]}}} \leftarrow \text{p-nitraniline}$$

Red. **Cloth Red G** (By.)

amidoazobenzene → 1-naphthol-4-sulphonic acid.

Red. **Brilliant Crocein M** (C.) (other brands)

amidoazobenzene → 2-naphthol-6 . 8-disulphonic acid.

Scarlet Red. **Crocein Scarlet 3B** (By.)

amidoazobenzene-4-sulphonic acid
→ 2-naphthol-8-sulphonic acid.

Black. **Coomassie Wool Black S** (Lev.)

p-phenylene diamine → α-naphthylamine
→ 2-naphthol-3 . 6-disulphonic acid.

Red. **Milling Red G** (C.)

2-naphthol-6-sulphonic acid ← thioaniline
→ 2-naphthol-6-sulphonic acid.

See also **Palatine Black** and **Naphthylamine Black D**. Many commercial acid blacks are mixtures of this latter dyestuff and **Naphthol Blue Blacks.**

Very few trisazo and tetrakisazo acid dyes are made commercially, and therefore these classes do not merit treatment here.

Azo-Acid Mordant Dyestuffs.—There is nothing to prevent an acid dyestuff also possessing mordant dyeing properties, in virtue of the presence of the usual mordant groups, in addition to acid groups. Such a dyestuff offers many advantages, its dual nature allowing of simple dyeing from an acid bath, with subsequent mordanting in a separate or the same bath, or in certain cases dyeing with the mordant present from the start. The dyeing operation is thereby greatly shortened, brighter shades are often obtained, and by the last method matching to shade is rendered easier. In addition to the usual mordant groupings, i.e., ortho-dioxy or ortho-oxy carboxyl groups, it has been found that in many cases ortho-oxy-azo compounds have mordant dyeing properties. Hence o-amidophenols are largely used as first components, and also their derivatives, such as sulphonic acids, nitro, and chlor derivatives. Such bodies are usually susceptible to oxidation, giving quinone compounds readily, and ordinary treatment with sodium nitrite and hydrochloric acid is not suitable for diazotising them. It has been found (Sandmeyer) that diazotisation is almost normal in presence of a copper salt, e.g., $CuSO_4$, and in absence of mineral acid. The addition of zinc sulphate (K.), excess sodium chloride (B.), and organic acids (W. t. M.) are other methods of rendering diazotisation easy.

Coupling of ortho diazo phenols is often very slow and difficult. This is accelerated by working in strong solutions with strong alkalis such as caustic soda or lime. Acetylation of certain amido groups may be necessary in the second component to protect them in such coupling. Another method of expediting coupling (Ber. & C.) is to condense the amidophenol with a sulphochloride, e.g., p-toluene sulphochloride thus:

$$H_2N—R—OSO_2—C_6H_4—CH_3.$$

The diazo body obtained from such a sulphonic ether readily couples, and the sulph-aryl group is then removed by saponification with hot dilute alkalis.

If in a dyestuff the azo group is ortho to a substituted chlorine atom, or even in the case of naphthalene bodies to a sulphonic acid group, these electro-negative groups are rendered easy of removal with weak alkali or even sodium acetate in aqueous solution, being thereby replaced by the hydroxy group (B., 1901, pat.). This method of obtaining a mordant grouping does not involve the diazotisation of amido phenols. Thus 2.6-diamido-1-chlorobenzene-4-sulphonic acid may be tetrazotised or diazotised in both amido groups, and on treatment of the tetrazo body with sodium carbonate, the chlorine atom is replaced by a phenolic group.

$$Cl{-}N_2{-}C_6H_2(Cl)(SO_3H){-}N_2.Cl \rightarrow Cl.N_2{-}C_6H_2(OH)(SO_3H){-}N_2{-}Cl$$

The tetrazo-oxy-sulphonic acid may be used for coupling.

In all cases mordanting causes more or less change in the shades given by these dyes although in certain cases the deepening may be small. In others the change in colour is considerable, e.g., **Chromotrope 2R** (M.) dyes wool red from an acid bath but on chroming subsequently becomes blue black. This dye is obtained by coupling diazobenzene with chromotropic acid.

$$C_6H_5.N_2{-}C_{10}H_2(OH)_2(SO_3Na)_2 \quad (\text{OH OH}; \ NaO_3S{-} \ldots SO_3Na)$$

Chromotrope 2B (M.) is obtained similarly by using

p-nitraniline, **6B** (M.) p-amido-acetanilide, **10B** (M.) α-naphthylamine, **8B** (M.) naphthionic acid, etc. all coupled with chromotropic acid (one molecule of each).

Other important azo acid-mordant dyes with the components from which they are derived are the following:

Metachrome Bordeaux B paste (Ber.)

picramic acid → m-phenylene diamine.

Chromazone Red A (G.)

p-amidobenzaldehyde

→ 1.8-dioxynaphthalene-3.6-disulphonic acid.

Acid Alizarin Brown B (M.), **Palatine Chrome Brown W** (B.)

o-amidophenol-p-sulphonic acid → m-phenylene diamine.

Acid Alizarin Violet N (M.), **Palatine Chrome Violet** (B.)

o-amidophenol-p-sulphonic acid → β-naphthol.

Diamond Black PV (By.)

o-amidophenol-p-sulphonic acid

→ 1.5-dioxynaphthalene.

Acid Alizarin Black R (M.)

6-nitro-2-amido-1-phenol-4-sulphonic acid → β-naphthol.

Palatine Chrome Black 6B (B.), also **Salicine Black U** (K.), **Eriochrome Blue Black R** (G.), **Acid Alizarine Blue Black A** (M.), **Diamond Blue Black EB** (By.), and other brands.

1-amido-2-naphthol-4-sulphonic acid → β-naphthol.

It is prepared by diazotisation of the first component by means of zinc nitrite with subsequent coupling in strong alkaline solution (coupling in presence of lime is often used for this class of body). An alternative method relies on preparing the first component from 1-amido-2.4-disulphonic acid by diazotisation and treatment with caustic soda whereby the 2-sulphonic acid group is replaced by hydroxyl.

Acid Alizarin Red B (M.)

anthranilic acid → 2-naphthol-3 . 6-disulphonic acid.

Anthracene Acid Brown G (C.)

sulphanilic acid → salicylic acid ← sulphanilic.

Cloth Red B (By.)

amidoazotoluene → 1-naphthol-4-sulphonic acid.

Acid Alizarin Black SE paste (M.)

β-naphthol ← 2 . 6-diamido-1-phenol-4-sulphonic acid
→ β-naphthol.

Anthracene Red (By.)

salicylic acid ← o-nitrobenzidine
→ 1-naphthol-4-sulphonic acid.

The **Mercerol** dyes (R.H.) belong to this class as do also the **Erachromes** (Lev.). The former contain their own mordant and dye like acid dyes. The **Erganon** dyes (B.), a new series used in calico printing, are obtained as soluble chrome compounds. They are fixed on the fibre by steaming or alkaline treatment when an insoluble chrome lake is fixed. The fastness is excellent.

Direct Cotton Dyes of the Azo Class. This class is a very large one and new members are still being added. Until **Congo Red** was discovered in 1883 by Böttiger no azo dyes were known which would dye cotton direct from aqueous solution. Benzidine and its homologues, when coupled to give dyestuffs, are found also to give cotton dyeing properties. The following diamines are used for this purpose :

benzidine $H_2N\langle\quad\rangle-\langle\quad\rangle NH_2$

tolidine $H_2N\langle\quad\rangle-\langle\quad\rangle NH_2$ (CH_3, CH_3)

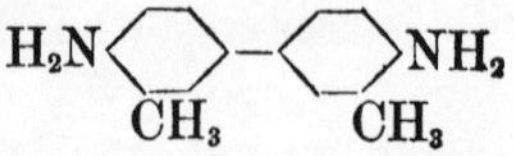

dianisidine $HN_2\langle\quad\rangle-\langle\quad\rangle NH_2$ (CH_3O, OCH_3)

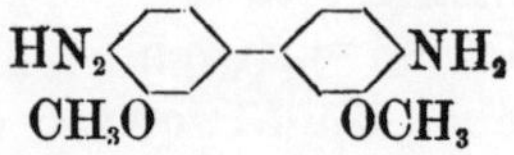

diphenetidine $H_2N\langle\quad\rangle-\langle\quad\rangle NH_2$ (C_2H_5O, OC_2H_5)

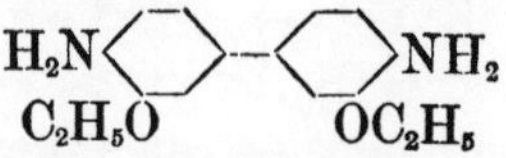

dichlorbenzidine $H_2N\langle\quad\rangle-\langle\quad\rangle NH_2$ (Cl, Cl)

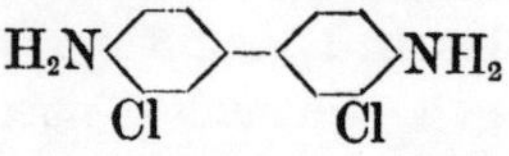

All the above ortho-substituted benzidine homologues have the property of furnishing azo dyes dyeing cotton direct, while meta-substituted derivatives have not this property or only to a much less extent. However, cotton dyeing properties are obtained where substitution gives a linkage in the meta position thus:

benzidine sulphone $H_2N\langle\quad\rangle-\langle\quad\rangle NH_2$ (bridged by SO_2)

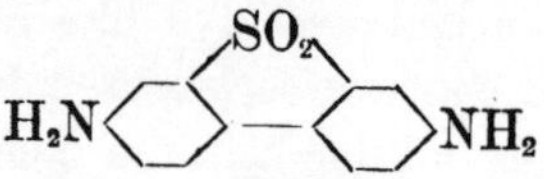

diamido-carbazol $H_2N\langle\quad\rangle-\langle\quad\rangle NH_2$ (bridged by NH)

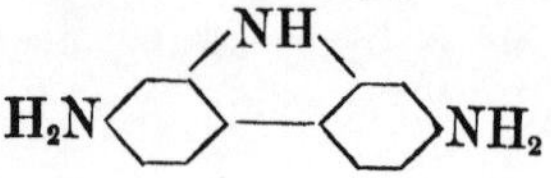

also diamido fluorene.

Besides the above diphenyl derivatives other para-diamines also give direct cotton-dyeing properties. Thus:

p-phenylene diamine $NH_2\langle\quad\rangle NH_2$

diamido stilbene $H_2N\langle\quad\rangle CH = CH\langle\quad\rangle NH_2$

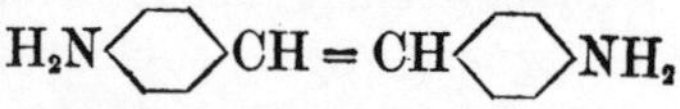

Also 4 . 4′-diamido-stilbene-2 . 2′-disulphonic acid.

p, p′-diamido diphenylamine

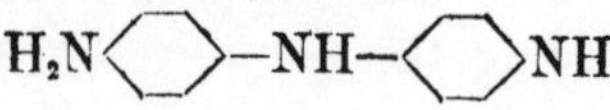

$H_2N\langle\quad\rangle-NH-\langle\quad\rangle NH_2$

p, p-diamido-diphenyl-urea

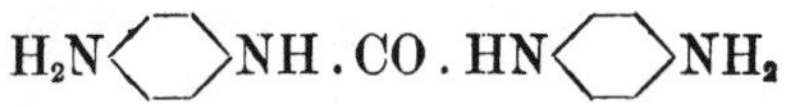

$$H_2N\langle\bigcirc\rangle NH.CO.HN\langle\bigcirc\rangle NH_2$$

Also a similar derivative of thio-urea $CS(NH_2)_2$.

Certain other para-diamines do not give cotton-dyeing properties to a derived azo dye, e.g., p, p′-diamido-dibenzyl.

Naphthalene diamine-1.5 and also its sulphonic acids like the above para-diamines give cotton-dyeing properties when diazotised and coupled to form azo dyes. Certain thiazol bases also give azo dyes having direct cotton-dyeing properties. (See p. 162.) On the other hand some substances which can be used as second components, namely, to which diazo bodies may be coupled, can also give direct dyeing properties to derived azo dyes.

Thus amido-naphthol sulphonic acid J is largely used for the above reason.

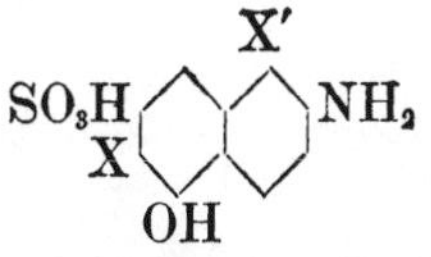

X′

SO_3H ⟨naphthalene ring⟩ NH_2

X

OH

X = point of coupling in alkaline solution.
X′ = ditto in acid solution.

There are certain peculiarities about J-acid and its azo derivatives. Thus coupled with diazobenzene derivatives cotton-dyeing properties are absent, but with diazonaphthalene derivatives are strongly marked even with some mon-azo dyes. By aryl- or acyl-substitution of the amido group still more affinity for cotton may be obtained, e.g., phenyl J-acid obtained by boiling with bisulphite and aniline.

The glycine derivative of J-acid obtained by treatment with chloracetic acid and other related derivatives are important (Lev.).

Again, the urea condensation product of this acid obtained by treatment with phosgene forms the second component of certain direct cotton dyes obtained by coupling on to it various diazo compounds.

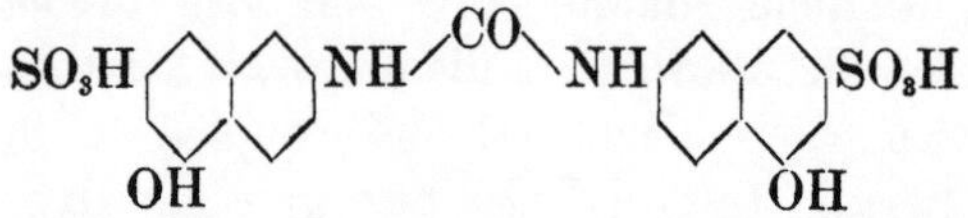

$$SO_3H \quad NH \overset{CO}{\diagup\diagdown} NH \quad SO_3H$$
$$OH \qquad\qquad\qquad\qquad OH$$

The **Benzo Fast Scarlets** (By.) are obtained from this body by further coupling, e.g., with sulphanilic acid, etc., ortho to the oxy groups.

Azidine Fast Scarlets (J.) are obtained from a still more complex urea derivative from 1-methyl-2 . 6-diamido benzene-4-sulphonic acid and 2 mols. J-acid. Toluidine or naphthylamine is coupled on this urea compound to give the dyes.

Dyes of this type are of good fastness to acids.

Certain strongly basic dyes possess direct cotton-dyeing properties and to this class belong the **Janus** dyes (M.), **Indoine Blue** and a few others. Such dyestuffs are faster however when applied with a tannin mordant or "back-tanned" after dyeing.

The direct cotton dyes dye from a neutral bath as salts, apparently the whole dyestuff molecule being taken up by the fibre, hence they are also known as salt dyes or substantive dyes. Many of them are also used on wool on account of their fastness to washing, rubbing and perspiration. The most marked characteristics of the class are a general tendency to stain white cotton effect-threads when soaped or milled, and a frequent sensitiveness to acids.

It has been found that by after-treatment, subsequent to dyeing, many of these dyes can be greatly improved in fastness. Certain of them are greatly improved towards washing by a weak chrome bath after dyeing, and others again towards light by a similar treatment with copper sulphate. Dyes with known mordant groupings invariably respond to such after-treatments as these, but in other cases also an improvement can frequently be distinguished. The presence of a hydroxyl group ortho to the azo group has been assigned as responsible for coppering taking place. After-treatment with formaldehyde is now extensively used for certain cotton dyes and great fastness to washing and milling is obtained in many cases. The action of formaldehyde is not yet clear but in many cases it seems to be connected with the presence of diamines or resorcin as an end component, although all such dyes are not susceptible to after-treatment. In the case of dyes having a resorcin component, Baekeland's theory of the formation of synthetic resins by action of formaldehyde on phenols is applicable. Probably a benzyl alcohol is first formed by the action of formaldehyde, this product next losing water in a condensation with another molecule of dye, the process repeating itself until a very stable complex is obtained. Special series of direct cotton dyes of this type have recently been put out, notably the **Vulcan** dyestuffs (Lev.), **Diamine Aldehyde** (C.), **Formal** (G.), **Benzoform** (By.), **Naphtoform** (K.), etc.

Other after-treatments include the use of "**Solidogen**" (M.), which is the methyl amido benzyl-p-toluide obtained by condensing formaldehyde with ortho- and para-toluidine, and also the use of sodium thiosulphate (H.) which increases light-fastness.

"Coupling" of the dye on the fibre after dyeing with diazotised p-nitraniline is sometimes employed. A new dye is thereby produced *in situ* which possesses a modified shade and fastness. The structure of the original dye must of course permit of coupling taking place readily.

In other cases "developing" is used, i.e., after dyeing, the goods are treated with nitrous acid to diazotise amido groups in such dyes, which can then be further "developed" by a bath of β-naphthol forming a more complex azo dye. In this way more complex bodies can actually be produced on the fibre than could satisfactorily be dyed after preparation in substance.

The **Diazo** series (By.) merits special mention in this connection. Certain direct cotton dyes offer unique advantages for application to wool when dyed as salt, acid or mordant dyes. In some cases they are put out commercially only as wool dyes.

Monazo Direct Cotton Dyes.

These are derived from thiazol bases, e.g., dehydrothio-p-toluidine diazotised and coupled with 1-naphthol-3.8-disulphonic acid gives **Erika 2GN** (Ber.), a red dyestuff.

OH SO_3Na

H_3C S C N_2—NaO_3S

N

The same first component with 1-naphthol-4.8-disulphonic acid gives **Geranine 2BG** (By.), with

1-naphthol-8-chlor-3 . 6-disulphonic acid **Diamine Rose R extra**, etc. (C.).

The homologous dehydrothiometaxylidine is used in **Erika B extra** and **G extra** (Ber.).

Other dyes are obtained by the use of Primuline sulphonic acid (see p. 179), thus **Dianil Yellow 3G** (M.) is obtained from this body, or dehydrothiotoluidine sulphonic acid, by diazotisation and coupling with aceto-acetic ether:

$$P . N_2 . CH\begin{cases} CO . CH_3 \\ CO_2 . C_2H_5 \end{cases}$$

Other dyes of this class with the second component required are as follows:

P = Primuline sulphonic acid,
D = dehydrothiotoluidine sulphonic acid.

Oriol Yellow (G.)
Cotton Yellow R (B.) } P (or D) → salicylic acid.

Clayton Yellow (Cl.), **Thiazol Yellow** (Ber., By.).
Titan Yellow G (R.H.), etc.

P (or D) → P (or D).

Rosophenine 10B (Cl.), **Thiazine Red R** (B.)

D → 1-naphthol-4-sulphonic acid.

Titan Pink 3B (R.H.)

D → 2-naphthol-6-sulphonic acid.

Thiazine Red G (B.)

P → 2-naphthol-6-sulphonic acid.

The above dyes are in general of good fastness to acids, especially the reds which are also fast to alkalis and soap in a moderate test.

Chloramine Yellow (By.), **Chlorophenine** (Cl.) is obtained by oxidation of 2 mols. of dehydrothiotoluidine sulphonic acid with hypochlorite. Four atoms of hydrogen are removed from the two amido groups giving an azo dye dyeing cotton very fast yellow shades.

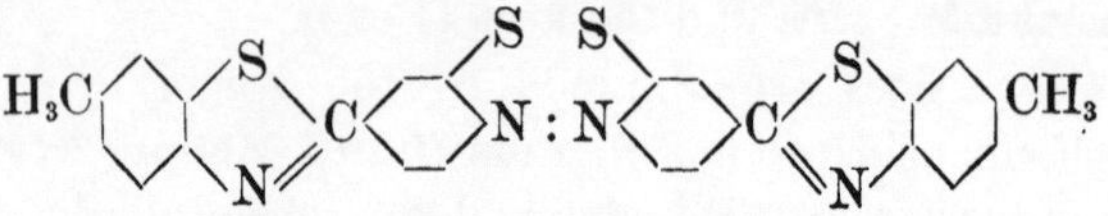

$$H_3C\text{—}C_6H_3\langle{}^{S}_{N}\rangle C\text{—}C_6H_3(S)\text{—}N:N\text{—}C_6H_3(S)\text{—}C\langle{}^{S}_{N}\rangle C_6H_3\text{—}CH_3$$

Disazo Direct Cotton Dyes.

Diaminogen Blue 2B (C.), **Diazanil Blue 2B** (M.)

1 . 4-naphthalene diamine-7-sulphonic acid
→ α-naphthylamine → 2-naphthol-6-sulphonic acid.

Dyes a deep blue, by diazotisation on the fibre and developing with β-naphthol gives a fast indigo shade.

Toluylene Yellow (G.E.)

1-toluylene-2 . 6-diamine-4-sulphonic acid ↗ 6-nitro-1 . 3-phenylene diamine
↘ 6-nitro-1.3-phenylene diamine.

The same diazo compound with only one molecule of m-phenylene diamine gives **Toluylene Brown G** (G.E.).

Diphenyl Fast Black (G.)

p . p′-diamidoditolylamine ↗ 6-toluylene-1 . 3-diamine
↘ 7-amido-1-naphthol-3-sulphonic acid
(alkaline coupling).

Benzo Fast Red 2BL (By.)

p . p′-diamidodiphenylurea ↗ 7-amido-1-naphthol-3-sulphonic acid
↘ ditto
(acid or neutral coupling).

Hessian Purple N (By., etc.)

4 . 4′-diamidostilbene-2 . 2′-disulphonic acid ↗ β-naphthylamine
↘ ditto.

Brilliant Yellow (By., etc.)

4 . 4′-diamidostilbene-2 . 2′-disulphonic acid $\begin{cases}\to \text{phenol} \\ \to \text{ditto.}\end{cases}$

Chrysophenine G, Pyramine Yellow G (B.), **Sultan Yellow G** (R.H.), etc., the most used of all cotton yellows is obtained from **Brilliant Yellow** above, by ethylation, e.g., with ethyl bromide, chloride, or sulphate.

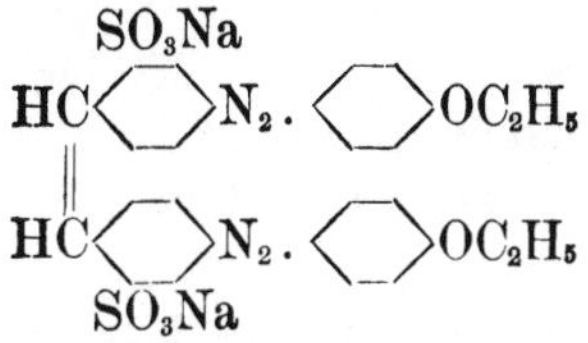

$$\begin{array}{l} \quad SO_3Na \\ HC\langle\bigcirc\rangle N_2 . \langle\bigcirc\rangle OC_2H_5 \\ \| \\ HC\langle\bigcirc\rangle N_2 . \langle\bigcirc\rangle OC_2H_5 \\ \quad SO_3Na \end{array}$$

This makes a dyestuff which is not sensitive to alkali, as is the case with **Brilliant Yellow.**

St Denis Red (P.), **Rosophenine 4B** (Cl.) obtained from diamidoazoxytoluene and Neville and Winther's acid has the formula:

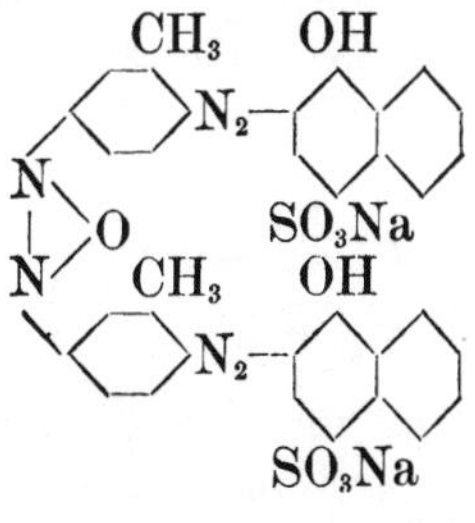

$$\begin{array}{l} \quad CH_3 \quad OH \\ \diagup\langle\bigcirc\rangle N_2-\bigcirc\bigcirc \\ N\diagdown \\ |\;\;\rangle O \quad\quad SO_3Na \\ N\diagup \quad CH_3 \quad OH \\ \diagdown\langle\bigcirc\rangle N_2-\bigcirc\bigcirc \\ \quad\quad\quad\quad SO_3Na \end{array}$$

It is dyed from alkaline baths and on washing the true red shade develops and is very fast.

Congo Red, the first direct cotton dye, is no longer

used extensively. It serves as an indicator for mineral acids which turn it blue. It has the following formula ·

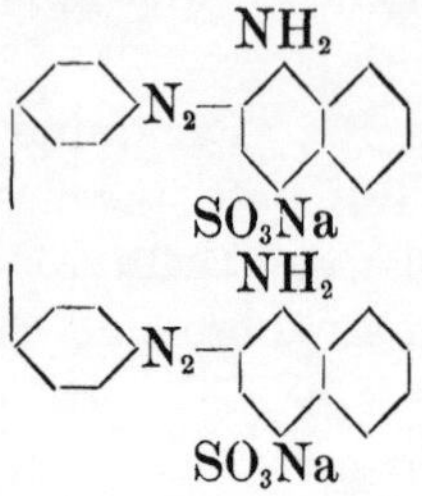

NH_2

N_2-

SO_3Na

NH_2

N_2-

SO_3Na

By use of tolidine instead of benzidine the superior **Benzopurpurine 4B** was obtained which is still extensively employed.

Other dyes of the benzidine type possessing special interest are detailed below. To some extent the shade can be predicted from the formulae of these dyestuffs, for example, **Chrysamine G** (By., etc.) is yellow.

benzidine ↗ salicylic acid
↘ ditto,

while by using one molecule of salicylic and one of naphthionic acid a shade mid-way between **Congo Red** and **Chrysamine G** is obtained, i.e., **Benzo Orange R** (By.)

benzidine ↗ 1-naphthylamine-4-sulphonic acid
↘ salicylic acid.

Benzidine and its homologues diazotise under ordinary conditions in both amido groups simultaneously. Coupling of the tetrazo bodies thus formed however takes place in two distinct stages, one diazo group coupling immediately and the other only after some time, e.g., in the manufacture of **Benzopurpurine 4B** the first stage is over in a few minutes out mechanical stirring and gradual addition of the sodium carbonate necessary for the second stage is continued for at least two days.

Another method (Badische) of obtaining disazo dyes of the benzidine type is to oxidise two simple azo compounds to get a diphenyl linkage. This method has little practical importance owing to the risk of destroying other groups.

Diazo Black B (By.)

benzidine ↗ 1-naphthylamine-5-sulphonic acid
benzidine ↘ ditto.

It is diazotised on the fibre and developed with β-naphthol for blue-black shades or m-phenylene diamine for brownish-blacks, or a combination of the two may be used.

Diamine Scarlet B (C.); the dyestuff is obtained from

benzidine ↗ 2-naphthol-6.8-disulphonic acid
benzidine ↘ phenol

by ethylation, which improves the alkali-fastness.

Diamine Black BH (C.) (many other brands), a developing cotton black of constitution

benzidine → (OH, NH_2, NaO_3S, SO_3Na naphthalene)
benzidine ↘ (OH, NH_2, NaO_3S naphthalene)

Diamine Blue 2B (C.) (many other brands)

benzidine ↗ 1.8-amidonaphthol-3.6-disulphonic acid
benzidine ↘ ditto.

Diamine Fast Red F (C.) (many other brands)

benzidine ↗ salicylic acid
↘ 2-amido-8-naphthol-6-sulphonic acid
(coupled first in acid solution).

Largely used on both wool and cotton, fastness improved in either case by after-treatment with bichrome or chromium fluoride.

Sulphonazurine D (By.), obtained from benzidine-sulphone-disulphonic acid and two molecules of phenyl-α-naphthylamine, is used on both cotton and wool giving blue shades of good fastness to milling, alkalis and acids. It has the following formula :

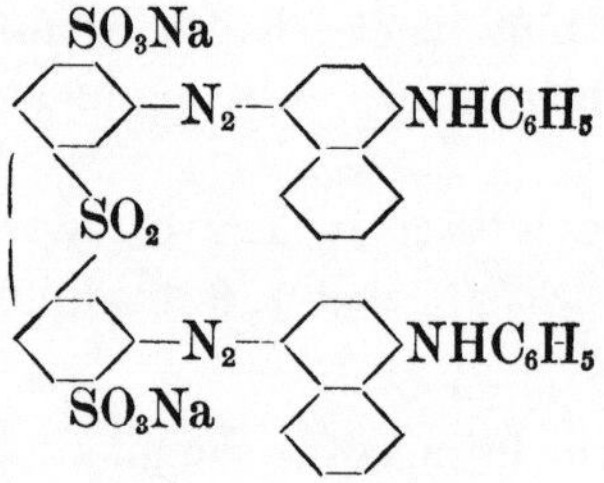

SO_3Na

$-N_2-$ NHC_6H_5

SO_2

$-N_2-$ NHC_6H_5

SO_3Na

Benzopurpurine 4B (many other brands), obtained from tolidine and naphthionic acid, has the formula

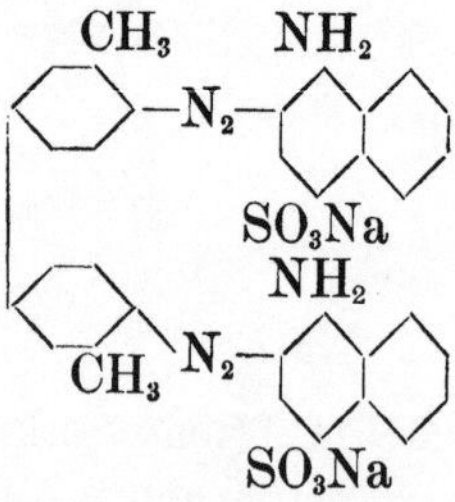

CH_3 NH_2

$-N_2-$

SO_3Na

NH_2

CH_3 N_2-

SO_3Na

The Benzopurpurines are all red dyestuffs.

Benzopurpurine 6B is isomeric with the above, being

tolidine ↗ 1 . 5-naphthylamine sulphonic acid
↘ ditto.

By diazotisation on the fibre and development with β-naphthol it couples further to give a fast black, **Diazo Brilliant Black B and R** (By.) is the same product applied by this method.

Benzopurpurine B,

tolidine ↗ 2-naphthylamine-6-sulphonic acid
↘ ditto.

Diamine Blue 3B (C.), obtained by alkaline coupling,

tolidine ↗ 1 . 8-amidonaphthol-3 . 6-disulphonic acid
↘ ditto.

It becomes redder with acids and violet with alkalis.

Benzopurpurine 10B, the bluest benzopurpurine red,

dianisidine ↗ naphthionic acid (1-NH_2-4-SO_3H)
↘ ditto.

The **Chicago Blues** (Ber.) are a series of blues of extraordinary brightness.

Chicago Blue 6B (Ber.) (other brands) is prepared from dianisidine and 1-amino-8-naphthol-2 . 4-disulphonic acid. The advantage of using this acid instead of H-acid (1 . 8—3 . 6) is in its constitution prohibiting any coupling on the amido side occurring as a side reaction. If such mixed coupling occurs, even to a small extent, as is often the case, the greatest brilliancy of shade cannot be expected from such a mixed product.

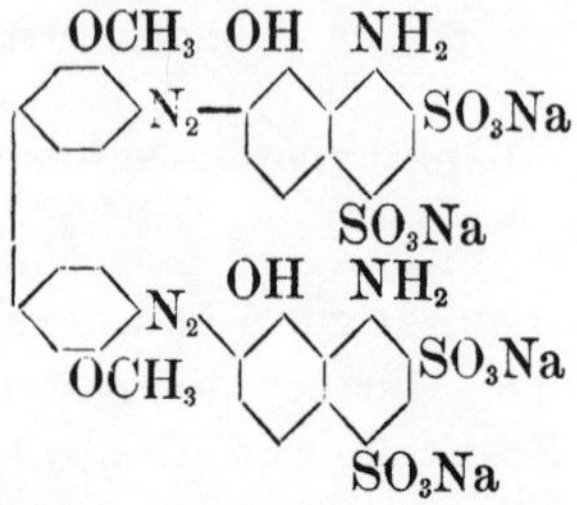

OCH₃ OH NH₂

$$\text{(structural formula: } C_6H_4(OCH_3)-N_2-C_{10}H_3(OH)(NH_2)(SO_3Na)_2 \text{ linked to } C_6H_3(OCH_3)-N_2-C_{10}H_3(OH)(NH_2)(SO_3Na)_2)$$

Diamine Gold (C.), the di-ethoxy-azo dye obtained by ethylation of the product from

1 . 5-naphthalene diamine-3 . 7-sulphonic acid ↗ phenol ↘ phenol.

Diamine Catechine (C.) is obtained by dyeing the disazo dye represented here

1 . 5-naphthalene diamine-3 . 7-sulphonic acid ↗ α-naphthylamine ↘ α-naphthylamine;

the violet shade thus obtained is **Naphthalene Violet** (C.) and on diazotisation and developing with warm soda solution gives a cutch-brown shade the amido groups in the second component (2 mols.) being replaced by hydroxy groups.

Trisazo Direct Cotton Dyes.

These supply the deeper shades of blue, also browns, greens, greys and blacks. Important typical examples are the following:

Columbia Black FF (Ber.), **Dianol Black FF** (Lev.), etc.

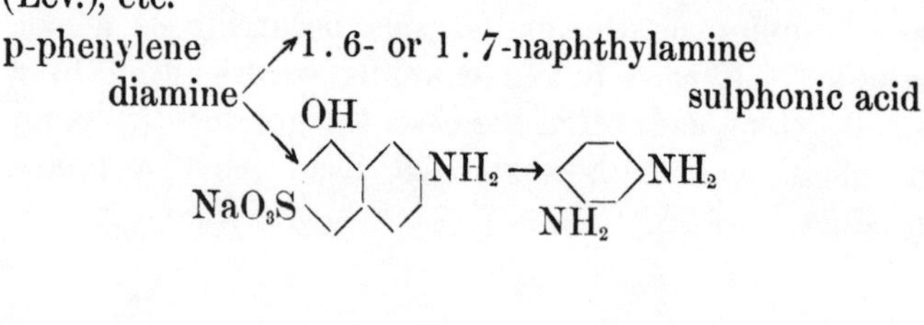

p-phenylene diamine ↗ 1 . 6- or 1 . 7-naphthylamine sulphonic acid ↘ $NaO_3S-C_{10}H_4(OH)-NH_2$ → $C_6H_3(NH_2)_2$ (OH, NH_2, NaO_3S; NH_2, NH_2)

Benzo Olive (By.).

benzidine ↗ salicylic acid
benzidine ↘ α-naphthylamine → 1 . 8-amido-naphthol-3 . 6-disulphonic acid.

Benzo Grey S extra (By.) is obtained by the use of the 1 . 4-naphthol sulphonic acid instead of the H-amido naphthol acid in the preceding example.

Diamine Green B (C.). The first green direct cotton dyestuff.

benzidine ↗ phenol
benzidine ↘ (OH, NH_2, NaO_3S, SO_3Na naphthalene nucleus) — N_2 — C_6H_4 — NO_2

Obtained by coupling para-nitraniline on to H-acid in acid solution, then coupling tetrazotised benzidine first with this azo compound and finally with phenol.

When salicylic acid is used instead of phenol **Diamine Green G** (C.) is obtained.

Tetrakrisazo Direct Cotton Dyes.

Such of these dyes as are in use are mainly browns and the class has a limited commercial importance.

Benzo Brown G (By.) is obtained by coupling two molecules of sulphanilic acid on to one of **Bismarck Brown.**

Toluylene Brown (G.E.) is obtained similarly by using naphthionic acid instead of sulphanilic.

Azo dyes of the anthracene series present greater difficulties in manufacture than those above described, among other causes being the insolubility of many intermediate products of this class, slow diazotisation and coupling. At present no azo-anthracene derivatives are of commercial importance as dyestuffs, but it is probable that in the near future this will no longer be the case.

Additional information may be acquired from a "Summary of Recent Progress in Colouring Matters," L. E. Vlies, *Journ. Soc. of Dyers*, p. 316, 1913, and pp. 22, 29, 1914, Thorpe's *Dict. of Applied Chemistry*, section "Azo Dyestuffs," Schultz, *Farbstoff-Tabellen*, and the patent record of Friedlaender, as well as abstracts of patents in the technical journals.

CHAPTER XVII

STILBENE, PYRAZOLONE AND THIAZOL DYESTUFFS

BELONGING to the azo class, or closely related to it, are the small groups of stilbene, pyrazolone and thiazol dyestuffs.

Stilbene Dyestuffs. This group comprises a number of yellow and orange dyestuffs dyeing cotton direct. On account of the methods of synthesis and constitution of these dyestuffs they are conveniently grouped together. They have the common property of giving shades fast to washing and alkali and often also to light and acids. The constitution of these bodies has only recently been fully elucidated. The methods used by Green in this task may be classed as (1) oxidation of the dyestuffs with permanganate to benzaldehyde sulphonic acids, (2) characterisation of oxidation products and (3) reduction of dyestuffs to diamido-stilbene disulphonic acid with titanous chloride.

In 1883 Walther obtained **Sun Yellow** (G.), also known as **Direct Yellow R**, **Curcumine S** and by many other names, by action of hot caustic soda solution on p-nitrotoluene-o-sulphonic acid. By oxidation of this dyestuff with hypochlorite a more greenish yellow **Mikado Yellow** (L.) is obtained, while alkaline reduction gives **Mikado Orange** (L.) and further reduction a leuco compound (obtainable from any of the abovementioned dyestuffs) which reoxidises in air to orange. Still further reduction gives diamidostilbene disulphonic acid. **Stilbene Yellows 8G** and **G** (B.) and **Stilbene Orange 4R** (Cl.) are obtained by alkaline reduction, e.g., with glucose and caustic soda, of dinitrostilbene disulphonic acid, during which successive reduction and condensation give the above dyestuffs, and subsequently further reduction gives a leuco orange, and ultimately diamidostilbene disulphonic acid. Of the identity of the corresponding **Mikado** (L.) dyestuffs and these **Stilbene** dyestuffs there is now no doubt.

The alkaline condensation of p-nitrotoluene sulphonic acid is largely influenced by the ortho electro negative substituent, and takes place in the following stages:

$$
\begin{array}{ccc}
\begin{array}{c} SO_3Na \\ H_3C\langle\quad\rangle NO_2 \\ \\ H_3C\langle\quad\rangle NO_2 \\ SO_3Na \end{array}
& -H_2O= &
\begin{array}{c} SO_3Na \\ H_2C\langle\quad\rangle NO \\ | \\ H_2C\langle\quad\rangle NO_2 \\ SO_3Na \end{array}
\end{array}
$$

Oxidation of this red intermediate compound gives dinitrodibenzyl disulphonic acid, from which **Mikado Yellow** (L.) may be obtained by heating with caustic soda solution. Rapid condensation, however, gives a blue intermediate body thus :

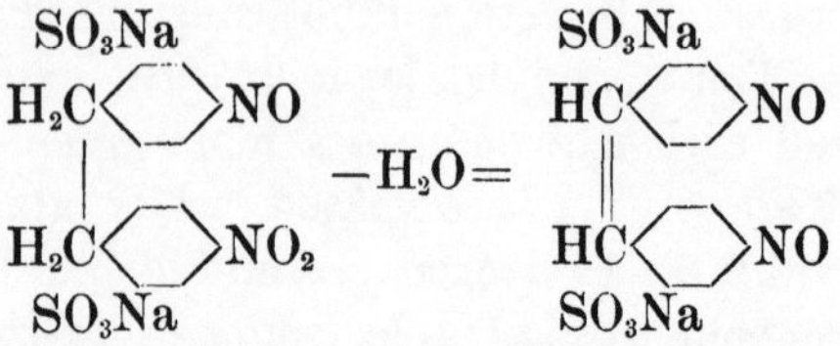

$$\begin{array}{c} SO_3Na \\ H_2C\langle\quad\rangle NO \\ | \\ H_2C\langle\quad\rangle NO_2 \\ SO_3Na \end{array} - H_2O = \begin{array}{c} SO_3Na \\ HC\langle\quad\rangle NO \\ \| \\ HC\langle\quad\rangle NO \\ SO_3Na \end{array}$$

Oxidation of this body readily gives dinitrostilbene disulphonic acid. In the ordinary preparation of **Direct Yellow R** or **G** (K.), the presence of aldehydic bodies formed simultaneously causes reduction, and by removal of an atom of oxygen two molecules of the dinitroso compound condense to give the dyestuff

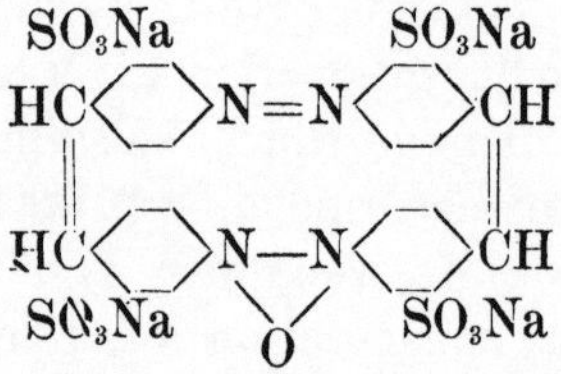

$$\begin{array}{ccc} SO_3Na & & SO_3Na \\ HC\langle\quad\rangle & N=N & \langle\quad\rangle CH \\ \| & & \| \\ HC\langle\quad\rangle & N-N & \langle\quad\rangle CH \\ SO_3Na & \diagdown O \diagup & SO_3Na \end{array}$$

This body on oxidation gives **Mikado Yellow,** identical with the **Stilbene Yellow 8G** obtained from dinitrostilbene disulphonic acid by alkaline reduction :

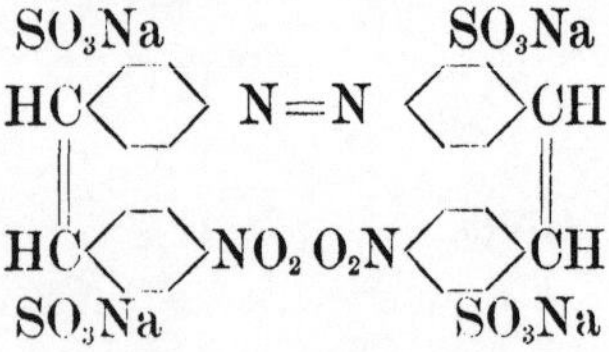

$$\begin{array}{ccc} SO_3Na & & SO_3Na \\ HC\langle\quad\rangle & N=N & \langle\quad\rangle CH \\ \| & & \| \\ HC\langle\quad\rangle NO_2 & & O_2N\langle\quad\rangle CH \\ SO_3Na & & SO_3Na \end{array}$$

Further reduction and condensation of the latter body give by a reverse change **Stilbene Yellow G** or **Direct Yellow R.** This azo-azoxy dyestuff by reduction gives **Mikado Orange** or **Stilbene Orange 4R** (Cl.), a disazo compound in which by further

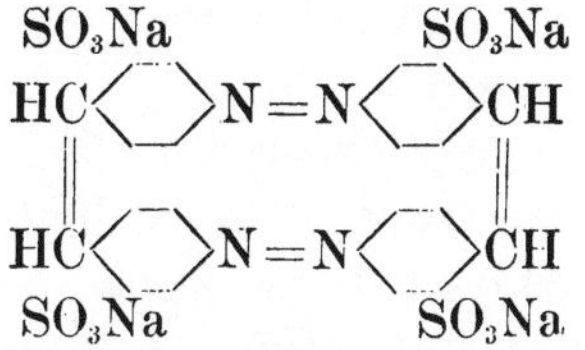

$$\begin{array}{ccc} SO_3Na & & SO_3Na \\ HC\langle\bigcirc\rangle N{=}N & & \langle\bigcirc\rangle CH \\ \| & & \| \\ HC\langle\bigcirc\rangle N{=}N & & \langle\bigcirc\rangle CH \\ SO_3Na & & SO_3Na \end{array}$$

reduction the azo groups become converted into hydrazo groups —NH—NH—, thus giving the leuco body which can reoxidise to orange in the air, while further reduction, as is usual with azo dyes, causes a dissolution of the azo groups to form amido groups, giving thus two molecules of colourless diamidostilbene disulphonic acid.

Another dyestuff of this series is **Diphenylcitronine G** (G.), obtained by alkaline condensation of dinitrodibenzyl disulphonic acid or dinitrostilbene disulphonic acid with two molecules of aniline, giving a disazo compound of the type $—N{=}N.C_6H_5$ Similarly, condensation of the intermediate dinitrosostilbene disulphonic acid above mentioned with p-phenylene diamine gives **Polychromine B** (G.), a brown cotton dyestuff.

Curcuphenine (Cl.), **Chlorophenine** and **Diphenyl Fast Yellow** (G.) are yellow stilbene dyestuffs obtained by condensations with dehydrothiotoluidine sulphonic acid.

Pyrazolone Dyestuffs. This group of dyestuffs, each of which contains the pyrazolone ring, have as

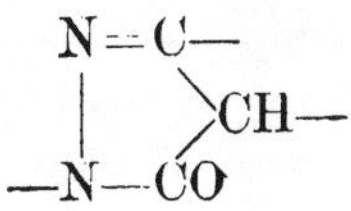

$$\begin{array}{l} \;\;N{=}C— \\ \;\;|\qquad \rangle CH— \\ —N—CO \end{array}$$

chromophore the azo group —N=N—. They belong to various dyeing classes, but are mostly yellows of good fastness to light, and are rather expensive products compared with many azo dyes.

The first to be discovered was **Tartrazine**, by Ziegler in 1884. It was obtained by the action of two molecules phenylhydrazine p-sulphonic acid on dioxytartaric acid in warm solution. Two molecules of water are first split off, giving a diphenyl-hydrazone thus:

```
                        CO₂H
                         |
                        C O H₂ N . NH . C₆H₄ . SO₃Na
                         |
NaO₃S . C₆H₄ . NH . N H₂ O C—COOH
```

A further splitting off of one molecule of water, thus:

```
        etc.
         |
         C—N . N H . C₆H₄ . SO₃Na
         |
    etc.=C—CO OH
```

results in the formation of a pyrazolone dyestuff

```
                       COOH (or Na)
                        |
                        C=N . N<C₆H₄>SO₃Na
                        |     /
NaO₃S<C₆H₄>NH . N=C—CO              Tartrazine.
```

An apparently more economical synthesis, involving the use of only one molecule of the expensive phenylhydrazine sulphonic acid, has since been accomplished. Oxal-acetic ester, condensed with phenylhydrazine sulphonic acid, gives a pyrazolone derivative with splitting off of water and alcohol, thus:

$$
\begin{array}{l}
CO_2C_2H_5 \\
| \\
CO \quad H_2N.NH C_6H_4.SO_3H \\
| \\
CH_2.CO\,OC_2H_5
\end{array}
$$

The second stage, i.e., the splitting off of alcohol, takes place easily, especially on heating. The sodium salt of the body so formed, treated with diazo sulphanilic acid or its anhydride $C_6H_4.N_2.SO_3$, gives an ester of **Tartrazine,** which is converted into the dyestuff itself by alkaline saponification.

$$
\begin{array}{rl}
 & CO_2C_2H_5 \\
 & | \\
 & C{=}N.N\langle\ \rangle SO_3Na \\
 & |\quad\ / \\
HO_3S\langle\ \rangle N:N.HC & -CO
\end{array}
$$

$$
\begin{array}{rl}
 & CO_2H \\
 & | \\
 & C{=}N.N\langle\ \rangle SO_3Na \\
 & |\quad\ / \\
NaO_3S\langle\ \rangle N:N.HC & -CO
\end{array}
$$

The azo dyestuff represented by this formula is apparently identical with the dyestuff as previously represented by a hydrazone structure. No conclusive evidence has been obtained which justifies the exclusion of either of these formulae. The case in point is an apt illustration of the inadequacy of simple benzenoid formulae for representation of dyestuffs without the assumption of tautomeric isomerism, which is part of the quinonoid theory, previously discussed.

Other pyrazolone dyestuffs are obtained by coupling diazo compounds with pyrazolone derivatives. For example, two direct cotton dyes, namely **Dianil**

Yellow R (M.) from **Primuline** and 1-phenyl-3-methyl-5-pyrazolone, and **Dianil Yellow 2R** (M.), by using instead of the latter body its para-sulphonic acid. In each case azo compounds are formed by coupling in position four of the pyrazolone ring.

Eriochrome Red B (G.) is an acid mordant dye obtained by coupling 1-amido-2-naphthol-4-sulphonic acid with phenylmethylpyrazolone.

Flavazine L & S (M.), **Fast Light Yellow G, 2G & 3G** (By.), **Hydrazine Yellow SO** (G.E.), **Xylene Yellow 3G,** and **Xylene Light Yellow 2G & R** (S.), are acid dyes fast to light, and are all obtained from simple diazobenzene derivatives and sulphophenyl-pyrazolones.

Thiazol Dyestuffs.—These dyestuffs have as chromophore the thiazol ring most simply illustrated in benzothiazol, itself colourless.

$$C_6H_4\left<\begin{matrix}N\\S\end{matrix}\right>C\,.\,C_6H_5$$

With the introduction of auxochrome groups into thiazols a yellow colour is obtained, as in dehydrothiotoluidine.

The thiazol dyestuffs are made by methods similar to that of Green, who discovered **Primuline** in 1887, i.e., by heating p-toluidine or its homologues with sulphur. According to the amount of sulphur used, the temperature and duration, different mixtures of thiazol bases are produced. Thus dehydrothiotoluidine is obtained in largest quantity by heating two molecules of base with four atoms of sulphur up to 200° C.

$$H_3C\cdot C_6H_4\cdot NH_2 + H_3C\cdot C_6H_4\cdot NH_2 + 4S = H_3C\cdot C_6H_3{<}{}^{S}_{N}{>}C\cdot C_6H_4\cdot NH_2 + 3H_2S$$

A considerable amount of primuline base is formed at the same time. The two bases can be separated by the difference of their solubility in alcohol, but they are used commercially as sulphonic acids, obtained by treatment of the powdered mixed bases with strongly fuming sulphuric acid, stirring well and cooling with cold water pipes and jacket. When complete, the acid mixture is run into water, and the sulphonic acid which precipitates from the acid solution is filtered off and washed. By treatment with strong ammonia, e.g., while still in the filter press, the ammonium salt of dehydrothiotoluidine sulphonic acid is not much dissolved, while that of **Primuline** readily dissolves. By adding salt to this solution, the commercial dye can be salted out as sodium salt.

In the manufacture of **Primuline** the melt is made in an iron pot with a long iron pipe sloping somewhat to the horizontal, this acting as air condenser for the toluidine. The sulphuretted hydrogen evolved may be piped back and burnt under the pot, the heat thus produced being sufficient to carry the reaction through once it is started. The temperature may be taken as high as 280° C. When the production of sulphuretted hydrogen ceases, the melt is run out, or blown out through a syphon tube with compressed air on a stone flag, where it sets to a hard yellowish brown mass.

Primuline contains both a di- and tri-thiazol derivative:

$$H_3C\langle\ \rangle\overset{S}{\underset{N=}{}}C\langle\ \rangle\overset{S}{\underset{N=}{}}C\langle\ \rangle NH_2 \text{ and}$$

$$H_3C\langle\ \rangle\overset{S}{\underset{N=}{}}C\langle\ \rangle\overset{S}{\underset{N=}{}}C\langle\ \rangle\overset{S}{\underset{N=}{}}C\langle\ \rangle NH_2$$

In sulphonating, the acid group enters the ring containing the NH_2 group.

Primuline dyes cotton a pure yellow shade, which is not fast to light. However, Green introduced the method of diazotisation on the fibre, and development with phenols or amines giving azo dyes of valuable properties, e.g., with β-naphthol, **Primuline Red.**

Thioflavine S (C.) is a methylated **Primuline,** and the replacement of $—NH_2$ by $—N(CH_3)_2$ renders the product less sensitive, although fastness to light still remains poor.

Thioflavine T (C.), also known as **Methylene Yellow H** (M.), **Rhoduline Yellow T** (By.), etc., is obtained from dehydrothiotoluidine by heating in an autoclave to 170° C. with methyl alcohol and sulphuric or hydrochloric acid. A basic ammonium substitution product is thereby formed, put out in commercial form as hydrochloride.

$$H_3C\langle\ \rangle\overset{S}{\underset{\underset{CH_3\ \ Cl}{N}}{}}C\langle\ \rangle N(CH_3)_2 . HCl$$

Thioflavine T is a pure greenish yellow, and is one of the most used basic yellows.

CHAPTER XVIII

DI- AND TRIPHENYLMETHANE DYESTUFFS

THE characteristic group which is also the chromophore in these two series is

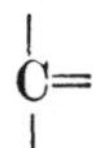

Oxidation of diphenylmethane $H_2C(C_6H_5)_2$ gives benzophenone $C_6H_5 . CO . C_6H_5$, which by reduction gives a secondary alcohol, i.e., diphenylcarbinol or benzhydrol C_6H_5—CH(OH)—C_6H_5, which by still further reduction gives diphenylmethane again.

In the case of triphenylmethane $HC(C_6H_5)_3$ oxidation gives the carbinol $HOC(C_6H_5)_3$. These bodies are all colourless, and to obtain colour it is necessary to have auxochrome groups in the para position to the methane carbon in bodies of the carbinol type, which then by a quinonoid change may become transformed to coloured isomers. Corresponding colourless and coloured isomeric carbinol and quinonoid bodies respectively are known. The ordinary benzene ring theory can only supply formulae inadequate to express these differences. The assumption of quinonoid change in the ring has been most fruitful in explaining such difficulties by supplying adequate formulae. From the point of view of the quinonoid theory of colour, the chromophore in the series under consideration is

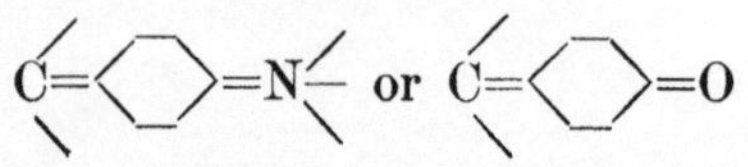

$$>C=\langle\ \rangle=N< \quad \text{or} \quad >C=\langle\ \rangle=O$$

according to the auxochrome groups in the para position. In certain cases it has not yet been definitely decided in which of the benzene rings substituted in the methane group the quinone structure ought to be written.

The only important diphenylmethane dyestuff is **Auramine O** (other brands). It was first obtained by Kern and Caro in 1883 by heating tetramethyldiamidobenzophenone (Michler's ketone) at 150° to 160° C. with ammonium chloride, and also zinc chloride, as dehydrating agent. Apparently water is split off and an imido group combined thus:

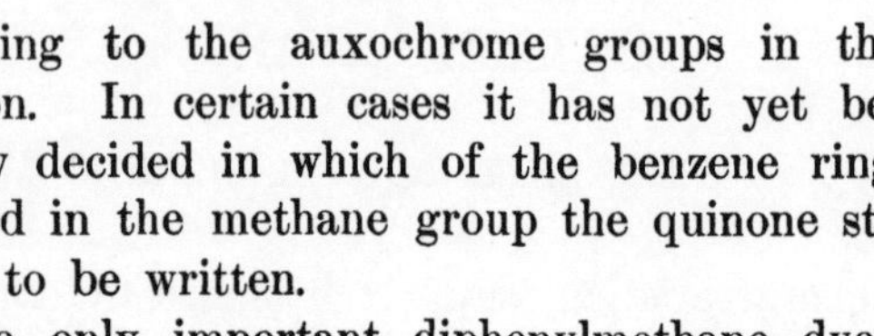

$$\begin{array}{c} C_6H_4N(CH_3)_2 \\ | \\ C\boxed{O\ H_2}NH\,.\,HCl \\ | \\ C_6H_4N(CH_3)_2 \end{array} \rightarrow \begin{array}{c} C_6H_4N(CH_3)_2\,.\,HCl \\ | \\ C=NH \\ | \\ C_6H_4N(CH_3)_2 \end{array}$$

Auramine base is colourless, but the hydrochloride shown above is a yellow basic dyestuff used in dyeing and printing on cotton (tannin mordanted), on wool and silk, and also for paper staining.

Auramine I, II, etc. are weaker brands liable to contain dextrine, which is the favourite adulterant for basic dyes.

The modern commercial synthesis of **Auramine** is by a different method. A melting-pot with upright pipe condenser (water-cooled) is heated in an oil bath, the contents of the pot are tetramethyldiamidodiphenylmethane, sulphur, ammonium chloride, and as diluent common salt. A stream of ammonia is passed during the course of the reaction. Ammonium sulphide is formed, the condensation giving the dyestuff as follows:

$$\begin{array}{l} C_6H_4 . N(CH_3)_2 \\ | \\ C\,H_2\ S\,S\ H_2\,NH \\ | \\ C_6H_4 . N(CH_3)_2 \end{array}$$

The melt is powdered-up, extracted with water and the dyestuff salted out.

As respresented above **Auramine** is a ketone-imide, and it behaves in this way on boiling with acids or even alone splitting off ammonia thus:

$$\begin{array}{l} | \\ C = NH \\ |\,O\ \ H_2 \end{array}$$

forming a colourless ketone. On this account the dye must not be applied above 60°—70° C.

Stock has shown that the phenyl derivative behaves as though a mono substituted amido group were present, namely,

$$\begin{array}{l} \| \\ C—NHC_6H_5 \\ | \end{array} \text{ and not } \begin{array}{l} | \\ C=NC_6H_5 \\ | \end{array}$$

On this basis the alternative quinonoid formula has been advanced.

$$\begin{array}{l} \quad /\!\!/ \langle\ \rangle = N(CH_3)_2 . Cl \\ C—NH_2 \\ \quad \backslash \langle\ \rangle N(CH_3)_2 \end{array}$$

Against this formula it may be urged that all known para quinonoid diphenylmethane derivatives are blue or violet.

Auramine G is obtained similarly by the initial use of methyl ortho toluidine instead of dimethylaniline, the ultimate product being an ortho (to the substituted amido groups) dimethyl derivative of **Auramine O.**

Triphenylmethane dyestuffs are very numerous, including basic, acid, mordant, and even direct cotton dyes. They are systematically classified as amido- and oxytriphenylmethanes, and it is more instructive to deal with them in this way than by a dyeing classification such as was adopted for the azo dyes. The pyronines, phthaleins and rhodamines, subsequently dealt with, are also, strictly speaking, of the triphenylmethane class.

Triamidotriphenylmethane Dyestuffs.—The oldest member of the triphenylmethane class is **Magenta** or **Fuchsin.** It is made commercially by heating a mixture of aniline with para and ortho toluidine in molecular proportions ("aniline for red") with an oxidising agent, e.g., arsenic acid or nitrobenzene with an iron salt as oxygen carrier. The latter agent is now preferred as giving a non-poisonous product. An excess of ortho toluidine is advantageous to the yield.

An enamelled iron melting-pot is used, heated by direct fire and fitted with stirring apparatus and in the lid a bent pipe and condenser, by which water and some excess oil, mainly o-toluidine and aniline, escapes. Apertures for removing test samples are fitted in the lid and a tube for running off the melt at the bottom.

Before heating the pot is charged with one part of "aniline for red," two parts of the dried hydrochloride of "aniline for red" and half a part of nitrobenzene. This is mixed with about 2—3 per cent. of iron filings during the heating which is continued up to 190° C. Test samples are removed to follow the course of the reaction, which is continued 12 hours after reaching 160° C.

A current of steam is finally blown in to remove by distillation the remaining uncombined oils, leaving a pasty mass of rosaniline base.

The pasty mass is dissolved in boiling hydrochloric acid solution to which salt is added salting out the crude **Magenta**, other bases dissolving. The bronzy mass is purified by dissolving in water under pressure. The insoluble products are filtered off.

Chalk or sodium carbonate may be added to precipitate azine bases and the resulting solution contains only rosaniline and chrysaniline. This is treated with hydrochloric acid to reform rosaniline hydrochloride or **Magenta**, and salt is added to aid crystallisation.

Crystallisation is done in large wooden tanks, in which are placed bars of wood or frames to aid the deposition of the large crystals, which take two or three days to grow. The yield of large crystals is 40—50 per cent. on the oils used.

The arsenic acid method offers great similarity, but gives a rather less yield.

The impurities of the magenta melt include acridines, e.g., chrysaniline or **Phosphine**, azines, e.g., **Indulines** and **Nigrosines**, and many other bodies. These are to a large extent worked up into marketable products: **Cerises** (B.), etc.

Constitution of Magenta. Nitrobenzene itself does not enter into the composition of **Magenta**, as is proved by use of chlornitrobenzene, which does not give a chlorinated dyestuff. All higher homologues of p-toluidine, in which meta and para positions are free, give dyes of the **Magenta** class.

By diazotisation of **Magenta** and boiling, an **Aurine** is obtained.

E. and O. Fischer in a classic investigation established the constitution of **Magenta** base or rosaniline base as a tri-p-amido-triphenylmethane.

The dyestuff obtained from p-toluidine and two molecules of aniline, $C_{19}H_{17}N_3 . HCl$, gave with alkalis the coloured base $C_{19}H_{17}N_3 . H_2O$ called pararosaniline, which by reduction gave a colourless or leuco base $C_{19}H_{19}N_3$. This leuco compound was shown to be a triamido compound, i.e., $C_{19}H_{13}(NH_2)_3$ by diazotisation and boiling with alcohol, whereby triphenylmethane itself was obtained.

$$HC \begin{cases} C_6H_5 \\ C_6H_5 \\ C_6H_5 \end{cases}$$

The usual **Magenta** obtained by oxidation of p-toluidine, aniline and o-toluidine under similar treatment gave homologous products, resulting finally in diphenyltolylmethane.

$$HC \begin{cases} C_6H_5 \\ C_6H_5 \\ C_6H_4 . CH_3 \end{cases}$$

The position of the substituent groups in **Magenta** was then determined by the inverse process of synthesis.

Nitration of triphenylmethane gave a trinitro derivative which on reduction gave a triamido compound identical with the leuco base, which by oxidation gave the coloured base of rosaniline, and this with hydrochloric acid gave the dyestuff itself.

$$HC\begin{cases} C_6H_4 . NO_2 \\ C_6H_4 . NO_2 \\ C_6H_4 . NO_2 \end{cases} \rightarrow HC\begin{cases} C_6H_4 . NH_2 \\ C_6H_4 . NH_2 \\ C_6H_4 . NH_2 \end{cases}$$

$$\rightarrow HO . C\begin{cases} C_6H_4 . NH_2 \\ C_6H_4 . NH_2 \\ C_6H_4 . NH_2 \end{cases} \rightarrow Cl . C\begin{cases} C_6H_4 . NH_2 \\ C_6H_4 . NH_2 \\ C_6H_4 . NH_2 \end{cases}$$

(old formula)

The position of substituents was established first from the fact that para toluidine or a para homologue is essential to get a rosaniline by the oxidation process, hence one amido group in **Magenta** is para to the methane carbon. This was also clear from the synthesis of **Magenta** by a condensation of p-amido-benzaldehyde with two molecules of aniline, heating with zinc chloride as dehydrating agent.

$$HC\begin{cases} C_6H_4 . NH_2 \\ \langle\bigcirc\rangle NH_2 \\ C_6H_4 . NH_2 \end{cases} \quad \text{leuco base}$$

In a similar condensation, using benzaldehyde instead of the amido derivative, diamidotriphenylmethane was obtained, which by diazotisation and boiling gave the corresponding di-oxy derivative.

$$HC\begin{cases} C_6H_4 . OH \\ . C_6H_5 \\ C_6H_4 . OH \end{cases} \rightarrow CO\begin{cases} \langle\bigcirc\rangle OH \\ \langle\bigcirc\rangle OH \end{cases}$$

Caustic potash fusion causes this di-oxy body to form the p-di-oxybenzophenone shown above. Hence it is deduced that the two amido groups supplied to rosaniline by aniline are also para to the methane carbon. The simplest rosaniline must be represented therefore as follows:

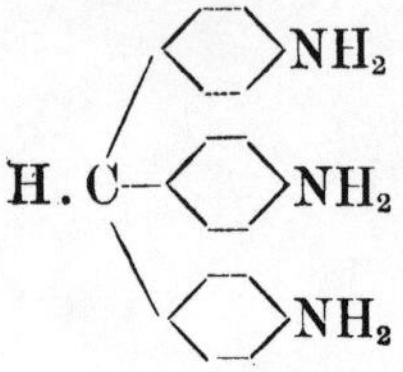

NH_2

H.C– NH_2

NH_2

leuco base

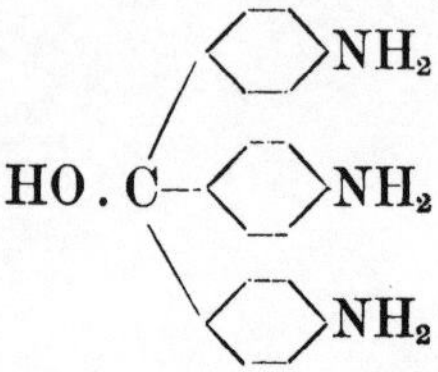

NH_2

HO.C– NH_2

NH_2

rosaniline base

NH_2

Cl.C– NH_2

NH_2

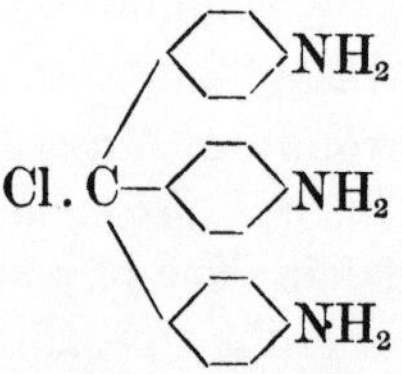

rosaniline hydrochloride

This p-rosaniline is present in commercial **Magenta** along with rosaniline, which is the tolyl homologue:

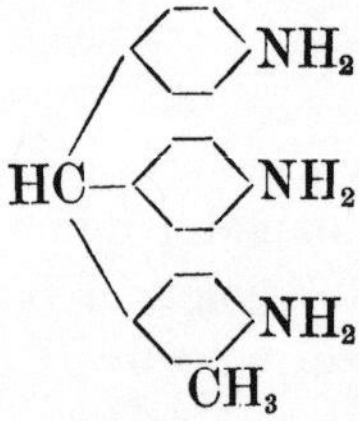

NH_2

HC– NH_2 leuco base

NH_2

CH_3

On the basis of the quinonoid theory of colour Rosenstiehl's formula for **Magenta** has been almost universally abandoned, and it is no longer written

$$Cl.C\begin{cases} -C_6H_4-NH_2 \\ -C_6H_4-NH_2 \\ -C_6H_4-NH_2 \end{cases} \quad \text{but instead} \quad C\begin{cases} -C_6H_4-NH_2 \\ =C_6H_4=NH_2Cl \\ -C_6H_4-NH_2 \end{cases}$$

Both coloured and colourless bases are known corresponding to the formulae

$$C\begin{cases} -C_6H_4-NH_2 \\ =C_6H_4=NH_2.OH \\ -C_6H_4-NH_3 \end{cases} \qquad HO.C\begin{cases} -C_6H_4-NH_2 \\ -C_6H_4-NH_2 \\ -C_6H_4-NH_2 \end{cases}$$

Hanztsch has found the former to exist in solution as a true ammonium base, whereas the colourless carbinol base is not dissociated. Colourless salts of the carbinol base have been prepared which readily change to coloured quinonoid isomers.

On heating rosaniline base, it loses one molecule of water, giving an anhydro base

$$C\begin{cases} -C_6H_4.NH_2 \\ =C_6H_4=NH \\ -C_6H_4.NH_2 \end{cases}$$

The formulae of rosaniline dyestuffs may be written as hydrochlorides of this base instead of chlorides of the quinonoid base.

$$C(C_6H_4NH_2)_2{=}C_6H_4{=}NH\,.\,HCl$$

Two general methods of synthesis of rosaniline dyestuffs have now been dealt with, i.e., (i) **the oxidation process**, (ii) **the substituted-benzaldehyde synthesis**, and the next to be dealt with is (iii) **the formaldehyde** or **New Fuchsin process.**

Formaldehyde and aniline in aqueous solution readily give anhydroformaldehyde-aniline (methylene aniline)

$$C_6H_5N{=}CH_2$$

which by heating with an excess of aniline and aniline hydrochloride undergoes isomeric change to anhydro-p-amidobenzyl alcohol, which readily condenses with another molecule of aniline on heating giving di-amidodiphenylmethane.

$$HN{<}C_6H_4{>}CH_2 + C_6H_5NH_2 = H_2C(C_6H_4NH_2)_2$$

By oxidation with nitrobenzene and ferric chloride as carrier, one molecule of aniline or o-toluidine may be condensed with this body to give homo- or para-rosaniline.

New Magenta obtained by this process has the formula

$$\mathrm{C}\begin{cases}\overset{\mathrm{CH_3}}{\langle\bigcirc\rangle}\mathrm{NH_2} \\ =\overset{\mathrm{CH_3}}{\langle\bigcirc\rangle}=\mathrm{NH\,.\,HCl} \\ \overset{\mathrm{CH_3}}{\langle\bigcirc\rangle}\mathrm{NH_2}\end{cases}$$

Other syntheses are by use of p-nitrobenzaldehyde with aniline (2 mols.) followed by reduction, while it is easily seen how rosaniline may be obtained by use of para-nitrobenzyl chloride and other benzyl derivatives.

The quinonoid theory may be used to explain the separate stages in condensations of the type dealt with. For example, by successive oxidation from benzenoid to quinonoid types, by which ready condensation with another molecule of an amine occurs, causing a higher phenylated methane to be formed:

$$\mathrm{H_3C}\langle\bigcirc\rangle\mathrm{NH_2} \rightarrow \mathrm{H_2C}=\langle\bigcirc\rangle=\mathrm{NH} + \langle\bigcirc\rangle\mathrm{NH_2}$$

$$\rightarrow \mathrm{H_2C}\begin{cases}\langle\bigcirc\rangle\mathrm{NH_2} \\ \langle\bigcirc\rangle\mathrm{NH_2}\end{cases}$$

which again by oxidation gives a quinone derivative leading to a further addition of aniline, thus forming the leuco base of rosaniline.

Another means of synthesis is used commercially in production of triphenylmethane dyestuffs, (iv) **the phosgene process.** Phosgene readily condenses with tertiary amines, e.g., with dimethylaniline. The gas is passed into the dimethylaniline. ($COCl_2$ is made by passing a mixture of chlorine and carbon monoxide over a catalyst, such as animal charcoal.) In this way tetramethyldiamido-benzophenone is formed by splitting off HCl.

$$(CH_3)_2N\langle\bigcirc\rangle\boxed{H\ Cl}.CO.\boxed{Cl\ H}\langle\bigcirc\rangle N(CH_3)_2$$

This type of benzophenone derivative on heating with another molecule of a tertiary amine in presence of phosphorus oxychloride, or by further treatment with phosgene, gives a triphenylmethane dyestuff, e.g., **Crystal Violet.**

$$Cl.C\begin{cases}\langle\bigcirc\rangle N(CH_3)_2\\ .\boxed{Cl\ H}\langle\bigcirc\rangle N(CH_3)_2\\ \langle\bigcirc\rangle N(CH_3)_2\end{cases} \rightarrow C\begin{cases}\langle\bigcirc\rangle N(CH_3)_2\\ =\langle\bigcirc\rangle =N(CH_3)_2Cl\\ \langle\bigcirc\rangle N(CH_3)_2\end{cases}$$

(v) Another method of synthesis is by condensation of tetra-acyl-diamido-benzhydrols with tertiary bases. This is really associated with the New Fuchsin process, in which the oxidation is carried through direct from the diamido-diphenyl-methane. By method (v) this latter body is oxidised in the cold to the alcohol or benzhydrol with lead peroxide and hydrochloric acid, and the benzhydrol so obtained condensed with another molecule of an amine, e.g., by heating in glacial acetic acid solution.

$$H_2C\begin{cases}C_6H_4.NR_2\\ C_6H_4.NR_2\end{cases} + C_6H_5.NR_2 = H.C(C_6H_4.NR_2)_3 + H_2O$$

Colour and General Properties of Triphenylmethane Dyestuffs.—The unsubstituted tri-amido compounds are red, substitution in the benzene ring having little effect upon shade. Substitution in the amido group causes a deepening of colour, e.g., hexamethylrosaniline hydrochloride (**Crystal Violet**) is violet, while triphenylrosaniline hydrochloride is blue (**Aniline Blue**). Removal of one para amido group giving diamido-triphenylmethane dyestuffs causes a change of shade to green or greenish blues; compare **Malachite Green**, etc., described later.

As explained above, reduction causes formation of colourless leuco compounds, which by slow re-oxidation in air, or quickly with an oxidising agent, give a return of colour. This is used as a test, employing hydrosulphite formaldehydes for reduction and very dilute persulphate for re-oxidation.

Oxidising agents readily destroy triphenylmethane dyes, ultimately giving quinone, and leuco compounds must be carefully oxidised to give dyes without waste, e.g. with theoretical amounts of lead peroxide. Closely connected with this question is that of the fastness of these dyes to light, which is not very great.

With excess of mineral acids, e.g., cold concentrated sulphuric acid, these dyestuffs give yellow solutions, which on dilution pass through mixed shades, finally giving the true colour on extreme dilution. This may be used as a confirmatory test in qualitative analysis.

Alkalis decolorise triphenylmethane dyestuffs, finally regenerating the carbinol bases with intermediate formation of a coloured ammonium quinonoid base (see above). The fastness to alkalis when dyed is somewhat limited for a similar reason.

Other important members of the triphenylmethane class are the following:

Methyl Violet—pentamethylrosaniline.

$$C\begin{cases} =C_6H_4=N(CH_3)_2Cl \\ -C_6H_4N(CH_3)_2 \\ -C_6H_4NH(CH_3) \end{cases}$$

This basic dyestuff is now obtained without the use of methyl iodide on rosaniline, by oxidation of dimethylaniline with air, using cupric chloride as carrier. One of the methyl groups leaves the nitrogen to give the central methane carbon. This probably involves intermediate formaldehyde formation and condensation to give the dyestuff.

The operation is carried out in large drums with loose lids to give free air exposure. Heating by means of a steam coil or jacket is used to start the reaction, which afterwards proceeds of itself.

Rotating stirrers give good mixing of the ingredients which include a large excess of common salt, also phenol or crude cresol, a small amount of water, and in some cases sand and an acid. The reaction giving the dyestuff takes place at 50°—60° C., when the mass takes on a shining metallic appearance. Common salt is apparently not merely a diluent, but a direct aid to the activity of the copper salt. The reaction takes about eight hours to complete. The melt is then treated with water and is purified by adding milk of lime, which precipitates the dye base and copper hydrate. These are filtered off and treated with sulphuretted hydrogen, e.g., by adding sodium sulphide afterwards acidifying with hydrochloric acid. The copper is precipitated and the dye dissolves as hydrochloride. The **Methyl Violet** solution is filtered off and salted out to obtain a purified product of which the yield is about 85 per cent.

By action of benzyl chloride on **Methyl Violet** the $—NH(CH_3)$ group is converted into $—N(CH_3).CH_2.C_6H_5$, giving a bluer dyestuff, i.e., **Methyl Violet 5B, 6B, and 7B,** or **Benzyl Violet.** The reaction is done under a reflux condenser in alcoholic solution in presence of a slight excess of alkali, at the boil for six to eight hours.

Ethyl Violet is the ethyl analogue of **Crystal Violet**, and is obtained by (i) condensation of diethylaniline with tetraethyldiamidobenzophenone, or (ii) action of phosgene on diethylaniline $(+ ZnCl_2)$, or (iii) oxidation of diethylaniline and tetraethyldiamidodiphenylmethane $(+ CuSO_4)$.

Methyl Green.

$$C\begin{cases} —C_6H_4—N(CH_3)_2 \\ =C_6H_4=N(CH_3)_2Cl + ZnCl_2 \\ —C_6H_4—N(CH_3)_2CH_3Cl \end{cases}$$

This strongly basic dye contains two substituted ammonium groups. It is obtained by treatment of **Methyl Violet** in amyl alcohol with methyl chloride.

Aniline Blue, Spirit Blue, Opal Blue—obtained by heating rosaniline base with excess aniline and benzoic or acetic acid under a reflux condenser to 180°.

Phenylation takes place according to the amount of aniline (or toluidine) used, giving phenylated derivatives. The benzoic acid is recovered from the melt unchanged. Its use is to expedite the reaction, but in what way is unknown. The di- and tri-phenylated para rosanilines are the most common of the dyestuffs thus prepared.

$$C{=}\left[\begin{array}{l} -C_6H_3(CH_3)NH(C_6H_5) \\ =C_6H_4{=}NH(C_6H_5)Cl \\ -C_6H_4NH(C_6H_5) \end{array}\right.$$

Victoria Blue 4R (B.), (Ber.), etc. is naphthylpentamethylrosaniline obtained by condensing methylphenyl-α-naphthylamine with tetramethyldiamidobenzophenone chloride.

Acid Magenta is an acid dye obtained by sulphonation of **Magenta** with fuming sulphuric acid (20 °/₀ SO_3), 120°—170° C., which gives a mixture of the di- and trisulphonic acids of homo- and para-rosaniline. The acid mixture is diluted, "limed out" and converted into the sodium salt by treatment with sodium carbonate.

$$OH{-}C\left[\begin{array}{l} -C_6H_2(CH_3)(NH_2)(SO_3Na) \\ -C_6H_3(NH_2)(SO_3Na) \\ -C_6H_3(NH_2)(SO_3Na) \end{array}\right. \qquad C\left[\begin{array}{l} -C_6H_2(CH_3)(NH_2)(SO_3Na) \\ -C_6H_3(NH_2)(SO_3Na) \\ =C_6H_3(SO_3){=}NH_2 \end{array}\right.$$

The coloured quinonoid inner anhydride is decolorised on treatment with dilute caustic soda, which converts it into the tri-sodium salt.

Acid Magenta is not very fast but is still extensively employed as a cheap easy levelling dye.

Acid Violet 4RS is the dimethyl derivative, i.e., with two —$NH(CH_3)$ groups, of **Acid Magenta.**

The **Acid Violets** are a numerous and important class of which one or two examples will suffice.

Acid Violet 4BN (B.), **6B** (By.), **N** (M.), etc. are obtained by sulphonation of **Benzyl Violet**, whereby tri-sulphonic acid is obtained. One sulphonic acid group enters the benzyl nucleus, otherwise it is analogous to **Acid Magenta.**

Formyl Violet S4B (C.) (many other **Acid Violet** brands) is prepared by a formaldehyde condensation of two molecules of ethylbenzylaniline sulphonic acid, the benzhydrol derivative thus obtained being further condensed with diethylaniline and oxidised from the leuco compound to give the dyestuff.

$$C\begin{cases} C_6H_4 . N(C_2H_5)CH_2 . C_6H_4 . SO_3 \\ =C_6H_4=N(C_2H_5)_2 \text{———————┘} \\ C_6H_4 . N(C_2H_5)CH_2 . C_6H_4 . SO_3Na \end{cases}$$

It is a valuable acid dye for animal fibres.

Sulphonated Aniline Blues.—By sulphonation of **Spirit Blue** mono-, di-, tri- or tetra-sulphonic acids may be obtained, soluble in water.

The mono-sulphonic acid is **Alkali Blue** (various brands), and this acid dye, insoluble as the free colour acid, is dyed from alkaline baths as the colourless salt, being developed blue on the fibre by a subsequent acid bath.

Soluble Blues and **Water Blues**, etc. are the mixed di- and tri-sulphonic acids and being more soluble can be dyed from acid baths.

These dyes are considerably used for cheap bright blues on animal fibres and tannin-mordanted vegetable fibres. The dyeing of the latter is dependent on the residual basic properties still potent in these acid dyes.

A somewhat similar type of body of unknown constitution is obtained by sulphonating the product from condensing diamido-diphenylmethane with rosaniline in presence of benzoic acid (R.H.). It dyes both animal fibres and unmordanted cotton. The most brilliant cotton blues have been obtained by dyeing a body of this type alone or in conjunction with benzidine derivatives. Thus the essential constituent of the **Chlorazol Brilliant Blues** (R.H.) is the di-sulphonate of tri-β-naphthyl-rosaniline.

The **Patent Blues** and **Cyanols** are sulphonated diamido-oxy-triphenyl methanes and in both their structure and greenish shade show a closer resemblance to the **Malachite Green** class of dyestuffs than to **Magenta**, but are faster to alkalis than either. They are most extensively used, being easy levelling acid dyes of good general fastness.

The condensation product of m-nitrobenzaldehyde with two molecules of diethyl or ethylbenzyl aniline gives a meta-nitro triphenyl methane body. The nitro group is then reduced to an amido group, diazotised and boiled with dilute acid to replace it by OH. The resulting m-oxy substituted diamido-triphenyl-methane is then disulphonated.

In preparing the **Cyanols** m-oxybenzaldehyde is used

initially but otherwise resemble **Patent Blue** in their general preparation and properties.

Patent Blue A (M.) (carbinol formula):

$$HO-C\begin{cases} -C_6H_4-N(C_2H_5).CH_2.C_6H_5 \\ -C_6H_4-N(C_2H_5).CH_2.C_6H_5 \\ -C_6H_2(OH)(HO_3S)-SO_3(\tfrac{1}{2}Ca) \end{cases}$$

Chrome Violet in paste (By.) is a mordant dyestuff obtained by oxidation of the product from condensing salicylic acid with tetramethyl diamido benzhydrol.

$$HO.C\begin{cases} -C_6H_4-N(CH_3)_2 \\ -C_6H_4-N(CH_3)_2 \\ -C_6H_3(CO_2H)-OH \end{cases} \quad \text{(carbinol formula)}$$

Malachite Green is obtained by heating 24 hours at 100° C. under reflux condenser in an enamelled stirring pan a mixture of one molecule benzaldehyde and two molecules dimethylaniline with a quantity of hydrochloric acid, insufficient to neutralise the base. (Zinc chloride was formerly used.) The mass is treated with caustic soda solution, the excess of base and benzaldehyde blown over with steam, and the oil remaining run into cold water where it solidifies while being well washed. It is then almost colourless and on oxidation in acid solution with the theoretical amount of lead peroxide the colourless leuco base gives the coloured carbinol. The lead is precipitated as sulphate, and the green solution of the coloured base as hydrochloride is decanted off, the dye salted out, or crystallised as a double chloride with $2ZnCl_2$. (Sometimes also it is prepared as oxalate with $3C_2H_2O_4$.) Further purification may be done by reprecipitation of the base with ammonia and retransformation into salt form.

$$=N(CH_3)_2 . Cl$$

$$C— \quad N(CH_3)_2$$

Particularly large crystals, measuring several inches, are made for export to the East[1].

Malachite Green is a bluish green basic dye, dyeing silk, wool, jute and leather direct, and also tannin mordanted cotton. It is also used in paper staining and lake manufacture. The dye itself, and also when on the fibre, is rather sensitive to alkalis, being readily converted into the carbinol form. By introduction of an ortho-chlorine substituent in the unsubstituted benzene ring greater fastness to alkali is obtained, i.e., increased stability. (Compare the ortho $—SO_3H$ group in **Patent Blues** and **Cyanols**.) Thus **Setoglaucine O** (G.) obtained by a similar condensation using o-chlorbenzaldehyde has improved properties. Similar dyestuffs to the latter are **Setocyanine O** (G.), **Glacier Blue** (J.), **Night Green A** (t. M.), **Patent Green AGL** (M.), **Brilliant Milling Green B** (C.), etc.,

[1] It even pays in this trade to pack the crystals in cotton wool to preserve their shape intact.

the last three dyestuffs being sulphonated compounds, namely acid dyes.

Guinea Green B (Ber.) and **Light Green SF** (B.) also belong to this class. The latter is obtained by condensation of benzaldehyde with methylbenzylaniline (**bluish** brand) or ethylbenzylaniline (**yellowish**), trisulphonation of the product and oxidation. One sulphonic acid group enters each benzyl nucleus, the other enters the unsubstituted ring in the para position.

The **Erioglaucines** (G.) are dyes of closely similar constitution. **Erioglaucine A** (G.):

$$\begin{array}{l} \quad\quad\;\; \diagup \bigcirc \, N(C_2H_5).CH_2.C_6H_4.SO_3NH_4 \\ OH-C-C_6H_4.SO_3H(2) \\ \quad\quad\;\; \diagdown \bigcirc \, N(C_2H_5).CH_2.C_6H_4.SO_3NH_4 \end{array}$$

Oxy Derivatives of Triphenylmethane.—These furnish only a small number of commercial dyestuffs. The oldest is **Aurin** or **Rosolic Acid** (Yellow Corallin) discovered by Runge in 1834. It finds small use now, only for varnishes and in photography. It is obtained by heating ten parts of phenol with five parts of strong sulphuric acid and six parts oxalic acid at 120°—130° C. The oxalic acid supplies the methane carbon atom and other condensations using CCl_4, or formaldehyde may be arranged to give rosolic acid. It is obtained pure by diazotising p-rosaniline and boiling.

$$\begin{array}{l} \quad \diagup \bigcirc \, OH \\ C = \bigcirc = O \\ \quad \diagdown \bigcirc \, OH \end{array}$$

Phenolphthalein is a well-known indicator giving bluish red coloration with alkalis, it finds some use also as a drug. The constitution of phenolphthalein has been elucidated with some difficulty and the subject has been closely identified with the discussion on the quinonoid theory of colour. The body is obtained by heating phthalic anhydride with zinc chloride or sulphuric acid as dehydrating agent.

$$C_6H_4\begin{matrix}\diagup CO \diagdown \\ \diagdown CO \diagup\end{matrix}O + 2C_6H_5OH = H_2O$$

$$+ \begin{matrix} HO\langle C_6H_4 \rangle - C - \langle C_6H_4 \rangle OH \\ \quad | \quad \diagdown O \diagdown \\ C_6H_4 - CO \end{matrix}$$

It is dioxyphthalophenone, phthalophenone itself being obtained by condensation of phthalyl chloride with two molecules of benzene in presence of aluminium chloride. Although a triphenylmethane derivative, phenolphthalein is also a **phthalein** (see later), and these bodies are strongly coloured in alkaline solution, but are decolorised by the weakest acids, e.g., H_2CO_3. Reduction of phthaleins gives phthalines:

$$\begin{matrix} -C- \\ | \diagdown O \diagdown \\ C_6H_4 - CO \end{matrix} + H_2 = \begin{matrix} HC- \\ | \\ C_6H_4 . COOH \end{matrix}$$

The lactone ring is also broken with salt formation when phthaleins are treated with alkali. Instant dehydration occurs however (compare **Aurin**) with formation of a quinone type, which is coloured.

$$HO\text{-}C_6H_4\text{-}C(OH)(C_6H_4.COONa)\text{-}C_6H_4\text{-}OH$$

$$\rightarrow HO\text{-}C_6H_4\text{-}C(C_6H_4.COONa)=C_6H_4=O \quad + H_2O$$

Probably the free (OH) group in the above formula by replacement gives (ONa).

Excess of alkali decolorises phenolphthalein giving

$$NaO\text{-}C_6H_4\text{-}C(OH)(C_6H_4.COONa)\text{-}C_6H_4\text{-}ONa$$

Neutralisation of the cooled solution with acetic acid gives quantitative removal of two Na atoms, the body still remaining colourless.

$$HO\text{-}C_6H_4\text{-}C(OH)(C_6H_4.COONa)\text{-}C_6H_4\text{-}OH$$

On heating or allowing to stand this solution again becomes coloured by dehydration and re-formation of quinone type.

Further evidence of the reality of this tautomerism has been found in the existence of two series of ethers of phenolphthalein, coloured and colourless according as they possess quinonoid or lactone structure.

Chrome Violet (G.) is the sodium salt of the tricarboxylic acid of **Aurin**, prepared by the action of formaldehyde on salicylic acid in strong sulphuric acid solution.

Other oxy-triphenylmethanes include the following acid mordant dyes, **Eriochromazurol B** (G.), **Eriochromcyanine R** (G.) and **Chromazurol S** (G.). These dyes, applied with chrome, are considerably used for commercial indigo-navy shades on wool fabrics. **Eriochromcyanine R** is obtained by condensation of benzaldehyde ortho-sulphonic acid with ortho-cresotinic acid.

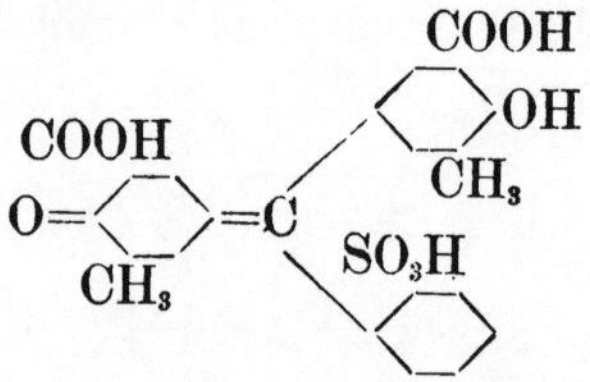

COOH
OH
COOH
CH_3
O= =C
SO_3H
CH_3

A few **naphthylphenylmethane** dyestuffs have attained commercial importance.

Victoria Blue B prepared by condensation of phenyl-α-naphthylamine with tetramethyldiamidobenzhydrol or tetramethyldiamidobenzophenone chloride.

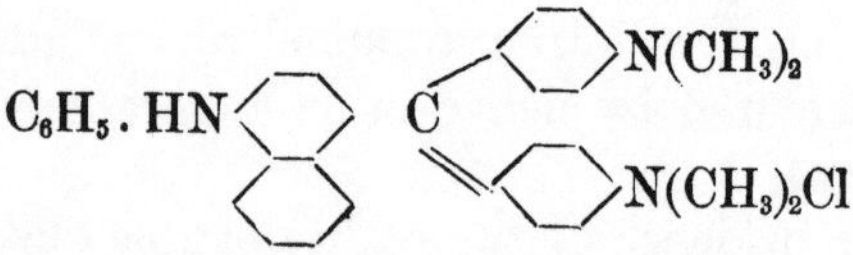

$N(CH_3)_2$
$C_6H_5 . HN$ C
$N(CH_3)_2Cl$

Victoria Blue R is prepared similarly using ethyl-α-naphthylamine.

In spite of being of inferior fastness to light, the **Victoria Blues** are very considerably used, for the brightest blues. They retain their purity considerably when viewed in artificial light.

Night Blue is of the same class, being prepared from tetraethyldiamidobenzophenone chloride and p-tolyl-α-naphthylamine.

Wool Green S (B.) is an acid dyestuff prepared by condensing β-naphthol with tetramethyldiamidobenzophenone chloride and sulphonating the product with fuming sulphuric acid.

Wool Green BS (By.) is obtained as above by use of G-salt instead of β-naphthol, and omitting also the final sulphonation.

CHAPTER XIX

XANTHENE DYESTUFFS

THE name of xanthene is applied to the anhydride of o-dioxy-diphenylmethane.

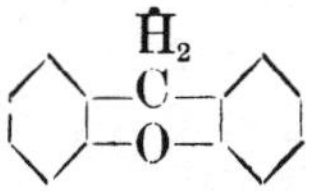

$$H_2$$
$$—C—$$
$$—O—$$

By oxidation it gives xanthone which by reduction gives a xanthhydrol which forms xanthonium salts with acids, the constitution of which demands the assumption of tetravalent oxygen (oxonium theory).

$$C_6H_4\begin{matrix} \diagup O \diagdown \\ \diagdown CO \diagup \end{matrix}C_6H_4 \rightarrow C_6H_4\begin{matrix} \diagup O \diagdown \\ \diagdown C \diagup \\ \diagup \ \ \diagdown \\ H \quad OH \end{matrix}C_6H_4$$

$$\begin{matrix} Cl \\ | \\ C_6H_4\begin{matrix} \diagup O \diagdown\!\!\diagdown \\ \diagdown C \diagup\!\!\diagup \end{matrix}C_6H_4 \\ H \end{matrix}$$

The xanthene dyestuffs offer great analogy with the di- and triphenylmethanes of which they are really derivatives. Thus the general method of preparation is to first prepare an ortho dioxy-p-disubstituted di- or triphenylmethane, to dehydrate forming the oxygen bridge, and to oxidise the leuco compound to give the dyestuff.

Fluorescein or **Uranin**, a phthalein, is obtained by heating resorcin with phthalic anhydride and conversion of the product into the sodium salt. The condensation takes place in two stages. First:

$$HO\langle\bigcirc\rangle OH \quad HO\langle\bigcirc\rangle OH \quad + \quad \begin{matrix} O \\ \| \\ C \diagdown \\ | \quad O \\ C_6H_4 - CO \end{matrix}$$

$$\rightarrow H_2O + \begin{matrix} HO\langle\bigcirc\rangle OH \ HO\langle\bigcirc\rangle OH \\ \diagdown - C - \diagup \\ | \ \diagdown O \\ C_6H_4 - CO \end{matrix}$$

This body then by dehydration gives a xanthene:

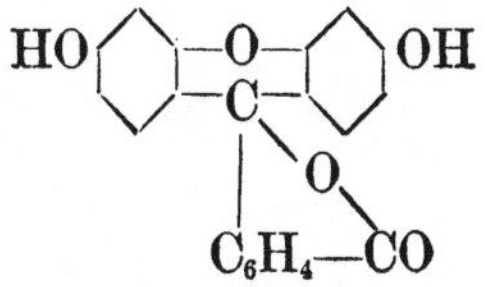

$$HO\,C_6H_3{-}O{-}C_6H_3\,OH \quad (\text{joined by } {-}C{-}, \text{ with } C{-}O{-}CO{-}C_6H_4{-}C)$$

which is colourless until treated with alkali when it forms a yellow dye soluble in water, showing a strong green fluorescence. The coloured sodium salt is represented by a quinonoid formula, namely

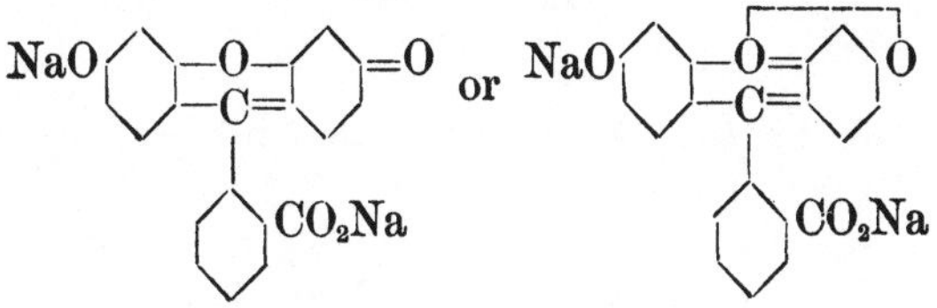

$$NaO\,C_6H_3{-}O{-}C_6H_3{=}O \;(\text{with } {-}C{=},\; C_6H_4\,CO_2Na) \quad \text{or} \quad NaO\,C_6H_3{-}O{=}C_6H_3\,O \;(\text{with } {-}C{=},\; C_6H_4\,CO_2Na)$$

The effect of alkalis on the lactone ring is similar to that in the case of phenolphthalein, and it has also been shown by Nietzki that isomeric coloured and colourless ethers exist. The dyestuff finds practically no use as such, but is important for preparation of the **Eosines.**

With the xanthene and phenylxanthene dyestuffs some doubt exists as to whether ortho or para quinone structure is more suitable. The oxonium structure finds growing favour.

The xanthenes may be subdivided into **pyrones** or **pyronines** and **phthaleins.** Closely allied to the latter are the small groups of **succineins** and **anthra-phthaleins.**

Pyronine Dyestuffs.—A small class possessing the pyrone ring.

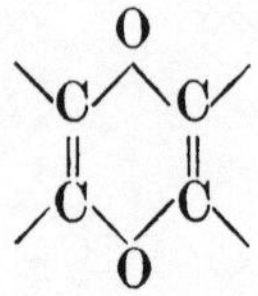

$$\begin{array}{c} O \\ \diagdown C \diagup \quad \diagdown C \diagup \\ \| \qquad \| \\ \diagup C \diagdown \quad \diagup C \diagdown \\ O \end{array}$$

They are red basic dyestuffs of minor importance obtained by formaldehyde condensation of substituted metamidophenols, e.g., dimethyl-m-amidophenol gives

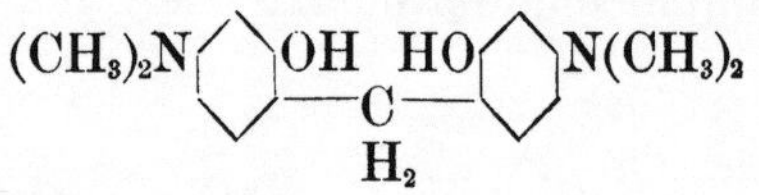

$$(CH_3)_2N-C_6H_3(OH)-CH_2-C_6H_3(OH)-N(CH_3)_2$$

which on heating with sulphuric acid is dehydrated, giving a leuco compound

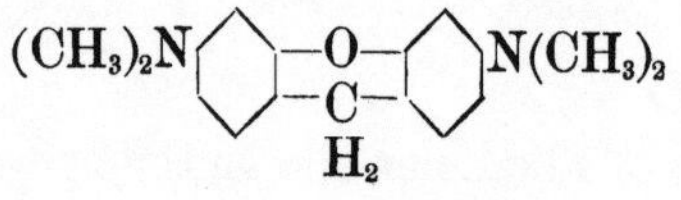

$$(CH_3)_2N-C_6H_3 \begin{array}{c} -O- \\ -CH_2- \end{array} C_6H_3-N(CH_3)_2$$

which by oxidation gives **Pyronine G**

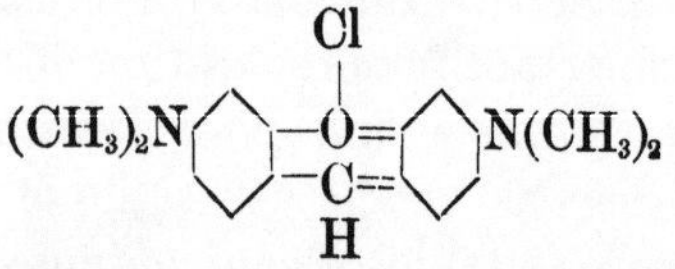

$$(CH_3)_2N-C_6H_3 \begin{array}{c} -O(Cl)= \\ -CH= \end{array} C_6H_3-N(CH_3)_2$$

Acridine Red B, 2B, 3B (L.) is obtained by oxidation of this body with permanganate when each $-N(CH_3)_2$ becomes oxidised to $-NH(CH_3)$.

Succinein Dyestuffs.—**Rhodamine S** is the only

important dye of this class; it is obtained by heating dimethyl-m-amidophenol with succinic anhydride.

$$(CH_3)_2N\!-\!C_6H_3\!\left\langle\begin{matrix} -\overset{\overset{Cl}{|}}{O}= \\ -\underset{\underset{C_2H_4-CO_2H}{|}}{C}= \end{matrix}\right\rangle\!C_6H_3\!-\!N(CH_3)_2$$

A para quinone formula may also be written.

Phthalein Dyestuffs.—The amidophthaleins or **Rhodamines** are bluish red basic dyes, mostly exhibiting fluorescence. They also have faintly acid properties in virtue of the possession of the phthalein carboxyl group, e.g., they do not precipitate so readily with tannin as most basic dyes. Acid dyes of this series are obtained by sulphonation.

The dyes of this class are derivatives of fluorane.

$$C_6H_4\!\left\langle\begin{matrix} O \\ C \end{matrix}\right\rangle\!C_6H_4 \qquad \begin{matrix} C-O \\ | \quad\;\; | \\ C_6H_4-CO \end{matrix}$$

Rhodamine B (B. and other makers) is obtained by condensation of phthalic anhydride with diethyl-m-amidophenol on heating.

$$(C_2H_5)_2N\!-\!C_6H_3\!\left\langle\begin{matrix} =\overset{\overset{Cl}{|}}{O}- \\ =\underset{\underset{C_6H_4-CO_2H}{|}}{C}- \end{matrix}\right\rangle\!C_6H_3\!-\!N(C_2H_5)_2$$

Rhodamine G is obtained from the above dyestuff by heating with aniline hydrochloride when one ethyl group is lost, and the shade becomes less bluish.

Rhodamine 3B is obtained from **Rhodamine B** by esterification with alcohol and HCl or by acting on the base with ethyl chloride. The shade is bluer than that of the **B brand**, and the fastness is increased by the aid to stability afforded by ester formation, which prevents ready return to the colourless lactone type. These ethylated **Rhodamines** are also known as **Anisolines** (Mon.) and are more basic than **Rhodamines**, having the property also of dyeing cotton direct.

Fast Acid Violet B (M.) (**Violamine B**) is obtained by the action of aniline or other primary aromatic amine on the di-chloride of **Fluorescein**. The reaction product is then sulphonated, giving

$$NaO_3SC_6H_4 . HN\langle C_6H_3 \rangle \begin{matrix} -O- \\ -C= \\ | \\ C_6H_4—CO_2H \end{matrix} \langle C_6H_3 \rangle =N—C_6H_5$$

When o-toluidine is used similarly, **Fast Acid Violet A2R** (M.) (**Violamine R**) is obtained.

Fast Acid Blue R (M.) is obtained when p-phenetidine replaces aniline above.

Fast Acid Eosine G (M.) is obtained by sulphonation of **Rhodamine.**

Intermediate between the amido- and oxy-phthaleins come amido oxy-phthaleins, partaking to some extent both of the character of **Rhodamine** and **Fluorescein**, being obtained by a combination of the methods of synthesis for these dyestuffs. These have been called **Rhodoles** but do not merit discussion here.

The oxy-phthaleins are derivatives of **Fluorescein** already described.

Chrysolin is obtained by condensation of resorcin, phthalic anhydride and benzyl chloride in presence of sulphuric acid.

NaO⟨ ⟩—O—⟨ ⟩=O
⟨ ⟩—C=⟨ ⟩$CH_2—C_6H_5$
|
$C_6H_4—CO_2Na$ (or o-quinone formula)

It is brown in solution with a green fluorescence. Halogenation of **Fluorescein** gives **Eosines**, red dyestuffs of greater fastness, being very fast to alkalis but only moderately to light. Fluorescence is still exhibited to a smaller degree.

The **Eosines** are weakly acid dyes and are applied from weakly acid baths to wool and silk.

Bromination of **Fluorescein** in aqueous or alcoholic solution and conversion of the tetra-brom-fluorescein into an alkali salt, gives **Eosine** or **Eosine G** (B.) (many other brands). In aqueous solution bromination is done with hypobromite.

Br Br
KO⟨ ⟩—O=⟨ ⟩O
Br⟨ ⟩—C=⟨ ⟩Br
|
$C_6H_4—CO_2K$

(A para quinone formula may also be written.)

The constitution of **Eosine** has been established from the following evidence. On heating with caustic soda it is broken down, giving dibromresorcin and dibromdioxybenzoyl-benzoic acid.

$$KO(Br)(Br)C_6H\langle{-O-}\rangle C\text{=}C_6H(Br)(Br)=O \;(\text{with } C_6H_4{-}CO_2K) \rightarrow KO(Br)(Br)C_6H{-}ONa \;\; {-}CO{-} \;\; NaO(Br)(Br)C_6H{-}ONa \;(\text{with } C_6H_4{-}CO_2K)$$

The free benzoic acid derivative by dehydration, e.g., with sulphuric acid, gives dibrom-xanthopurpurin according to a general reaction for preparation of anthraquinone derivatives.

$$(OH)_2(Br)_2C_6\langle{CO \atop CO}\rangle C_6H_4$$

Further, if the free benzoic acid compound be heated, the original **Eosine** is formed along with phthalic acid. The original **Eosine** is also re-formed when dibrom-resorcin is heated with phthalic acid. The constitution of **Eosine** is therefore established as a symmetrically brominated body shown in the above ortho-quinonoid formula or in the following p-quinone structure.

$$KO(Br)(Br)C_6H\langle{-O- \atop -C=}\rangle C_6H(Br)(Br)=O, \quad C{-}C_6H_4{-}CO_2K$$

Methylation of **Eosine** gives **Eosine spirit soluble**, i.e., instead of $—C_6H_4—CO_2K$ above, $—C_6H_4—CO_2CH_3$. It is also somewhat soluble in water as is **Eosine S** (B.), a similar product obtained by ethylation with alcohol and sulphuric acid.

Eosine BN (B.) is obtained by nitration of dibrom-fluorescein. This dye is one of the fastest of its class to light and milling. It gives bluish red shades and has only a weak fluorescence.

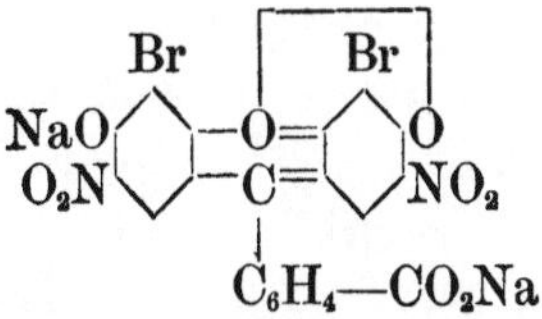

Br Br

NaO —O= O

O_2N —C= NO_2

C_6H_4—CO_2Na

By introducing iodine into **Fluorescein**, homologous dyestuffs of a more bluish character than **Eosine** are obtained, i.e., **Erythrosines.** The preparation is similar to that of **Eosine.**

Di-iodo-fluorescein is known as **Erythrosine extra yellowish** (B.) and tetra-iodo-fluorescein as the **extra bluish** brand.

By use of dichlor-phthalic acid instead of phthalic acid in manufacture of **Fluorescein**, a dichlor-fluorescein is obtained, which on bromination gives **Phloxine.**

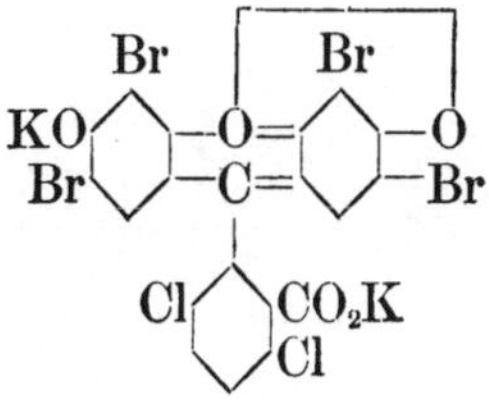

Br Br

KO —O= —O

Br —C= —Br

Cl CO_2K

Cl

(or a p-quinone formula)

Rose Bengale is the tetra-iodo derivative of the same dichlor-fluorescein.

Other brands of **Phloxine** and **Rose Bengale** are obtained from tetra-chlor-fluorescein derived from tetra-chlor-phthalic acid.

The **Cyanosines** are spirit soluble bodies used in lake manufacture and are obtained by alkylation of **Phloxines** (compare **Eosine spirit soluble**).

The only important mordant dye of the phthalein class is **Gallein** or **Alizarin Violet.** It is obtained by heating phthalic anhydride with gallic acid or pyrogallol; in either case the product is the same since CO_2 is split off from gallic acid during the heating at 190°—200° C.

```
      OH      OH
HO⟨  ⟩—O—⟨  ⟩=O
      —C=
       |
      C6H4—COOH
```

By heating **Gallein** with sulphuric acid to 200° C. a further molecule of water is split off giving **Coerulein** (B.), a green mordant dye put out as a paste, also as a water-soluble powder in the form of bisulphite compound $C_{20}H_{10}O_6 + NaHSO_3$, **Coerulein S.**

```
      OH      OH
HO⟨  ⟩—O—⟨  ⟩=O
      —C=
       |    |
      ⟨  ⟩—CO
```

(o-quinone formula also)

It will be seen that **Coerulein** is a derivative of anthraquinone, and it has recently been found to have vat-dyeing properties. Its commercial importance however is dependent on its use as a mordant dye.

Coerulein B, BR, BW, BWR (M.) are obtained by heating **Fluorescein** with excess sulphuric acid.

```
HO⟨  ⟩—O—⟨  ⟩=O
      —C=
       |    |
      ⟨  ⟩—CO
```

CHAPTER XX

ACRIDINE AND QUINOLINE DYESTUFFS

Acridine Dyestuffs.—These are allied to diphenylmethane and xanthene dyestuffs. Dihydroacridine is obtained by heating o-diamido-diphenylmethane, ammonia being split off, giving

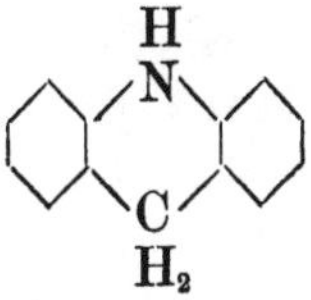

H
N
C
H_2

which on oxidation gives acridine which is represented by

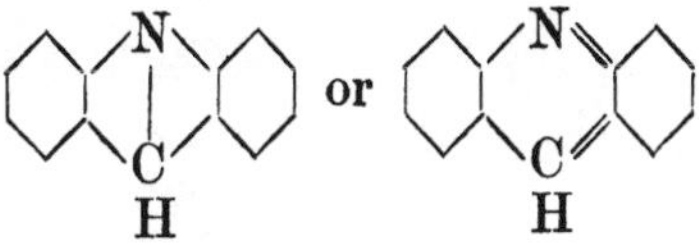

N C H or N C H

Acridine is found in coal tar also; it is a yellow basic body, fluorescent in solution. Phenylacridine is thus a derivative of triphenylmethane. The introduction of auxochrome groups para to the methane carbon in acridine gives yellow and orange dyestuffs mostly showing a greenish fluorescence in solution. Those of commercial importance are all basic.

Phosphine or **Leather Yellow** is obtained as a by-product of the **Magenta** melt. The residues from this process are reduced, the triphenylmethane bodies from stable leuco compounds; the **Phosphine** may be then separated as the difficultly soluble nitrate. The chief colouring matter has been called **Chrysaniline.**

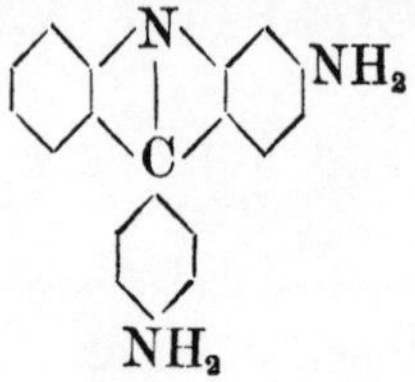

N, C, NH_2, NH_2

It is formed in **Magenta** manufacture by a partial ortho condensation instead of the main para condensation giving **Rosanilines.** On diazotisation and boiling with alcohol **Chrysaniline** gives phenylacridine. The dyestuff can be synthesised from o-nitro-benzaldehyde and two molecules of aniline by heating to get a triphenylmethane which is further reduced and oxidised.

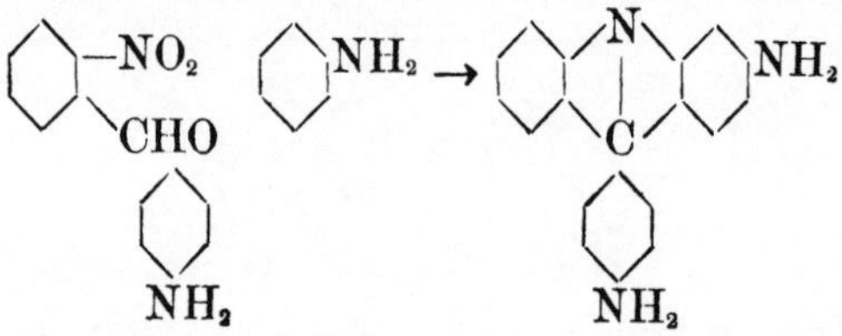

$-NO_2$, CHO, NH_2, NH_2 → N, C, NH_2, NH_2

The dye gives orange yellow solutions with green fluorescence and is used in cotton dyeing, printing, and in the dyeing of leather.

Acridine Orange R extra is obtained by condensation of benzaldehyde with m-amido-dimethylaniline in alcoholic hydrochloric acid solution. By heating the triphenylmethane derivative thus formed with hydrochloric acid under pressure, ammonia is split off and a nitrogen bridge formed also. The leuco compound is then oxidised to the dyestuff with ferric chloride.

$$(CH_3)_2N\text{—}C_6H_3\langle{}^{N}_{C(C_6H_5)}\rangle C_6H_3\text{—}N(CH_3)_2 . HCl$$

A similar condensation of benzaldehyde with m-tolylene diamine gives **Benzoflavin** (G.E.).

The above are phenylacridines, the simpler acridines have been since obtained by the use of formaldehyde. Thus **Acridine Yellow** is obtained by condensing formaldehyde with m-tolylene diamine.

$$\begin{matrix} H_2N \\ H_3C \end{matrix} C_6H_2\langle{}^{N}_{CH}\rangle C_6H_2 \begin{matrix} NH_2 . ClH \\ CH_3 \end{matrix}$$

When m-amido-dimethylaniline is used as above an **Acridine Orange NO** (L.) is obtained.

$$(CH_3)_2N\text{—}C_6H_3\langle{}^{N}_{CH}\rangle C_6H_3\text{—}N(CH_3)_2 . HCl$$

It is put out as a double zinc salt, i.e. + $ZnCl_2$.

Rheonine (B.), a brownish yellow dye with green fluorescence, of good fastness when dyed, is obtained from tetra-methyl-diamido-benzophenone by heating with m-phenylene diamine hydrochloride and zinc chloride.

In this way m-amido-phenylauramine (I) is formed intermediately by splitting off of water, and on further heating to 200° C. gives a leuco compound oxidising to the dyestuff (II).

$(CH_3)_2N$ — N — NH_2 — C — $N(CH_3)_2$ I

$(CH_3)_2N$ — N — NH_2 — C — $N(CH_3)_2$ II

Various **Patent Phosphines**, etc. are obtained by alkylating acridine dyestuffs, whereby acridinium compounds are formed which are strongly basic dyes, e.g., with methyl chloride and diamidoacridine.

$$H_2N{-}C_6H_3\langle{}^{N(Cl)(CH_3)}_{CH}\rangle C_6H_3{-}NH_2$$

Quinoline Dyestuffs.—A small group of which **Quinoline Yellow** is by far the most important. It is obtained from quinaldine by heating with phthalic anhydride.

$$C_6H_4\langle{}^{CO}_{CO}\rangle CH{-}C_9H_6N$$

It should be noted that water is split off with the other oxygen of phthalic anhydride to that involved in xanthene syntheses. The dye thus obtained is insoluble in water but soluble in spirit.

Quinoline Yellow S (or **water-soluble**) is obtained by sulphonation of the above dye with fuming sulphuric acid. The resulting disulphonic acid, containing some mono-sulphonic acid, has these acid groups in the quinoline residue. It is an acid dye giving the purest greenish yellow shades obtainable on textile fibres, possessing also good fastness to light. It is necessarily rather expensive.

Quinoline Red, obtained by action of benzotrichloride on raw quinoline (containing isoquinoline) along with zinc chloride, is used in colour photography, being sensitive to light, as is also **Quinoline Blue** or **Cyanin**, obtained by the action of caustic alkalis on the condensation product of amyliodide and a mixture of quinoline and lepidine (γ-methyl quinoline).

Pinacyanol (M.) and other **Pina-** dyes belong to this class.

(For the application of these dyes in photography see "Photography in Colour," by A. E. Woodhead, *Journ. Soc. Dyers*, p. 78, 1914.)

CHAPTER XXI

INDAMINES AND INDOPHENOLS

THESE derivatives of quinone imide comprise numerous blue and green dyestuffs which, while lacking the stability necessary for use as commercial dyestuffs, are valuable intermediate stages in the manufacture of other quinone imide dyestuffs, i.e., of the azine and related classes.

Quinone chlorimides have been obtained by oxidation of p-amidophenol with hypochlorites, while more recently Willstaetter has isolated the mono- and di-imides of quinone.

$$O=\langle\quad\rangle=NH \quad HN=\langle\quad\rangle=NH$$

The indophenols and indamines are derivatives of the above p-quinonoid bodies, the azine dyestuffs are derived from the corresponding o-quinonoid compounds.

Oxidation of aniline and p-phenylene diamine, e.g., with chromic acid, gives an indamine, **Phenylene Blue**:

$$H_2N\langle\quad\rangle N=\langle\quad\rangle=NH\,.\,HCl$$

Similarly oxidation of dimethyl-p-phenylene diamine and dimethyl aniline gives **Bindschedler's Green**:

$$(CH_3)_2N\langle\quad\rangle N=\langle\quad\rangle=N(CH_3)_2Cl$$

A second general method of synthesis is by acting on amines having a free p-position with nitrosamines:

$$(CH_3)_2N\langle\quad\rangle \boxed{H \; HO} N=\langle\quad\rangle=N(CH_3)_2Cl$$

If instead of a nitrosamine a nitrosophenol be taken, a similar reaction occurs with production of an indophenol. The indophenols are also obtained by oxidation of a p-diamine in presence of a phenol or naphthol having a free p-position.

Only one of these bodies has found commercial use, namely, **Indophenol Blue**:

$$(CH_3)_2N\langle\quad\rangle N=\langle\quad\rangle=O$$

This dyestuff has found a limited use as an addition to the indigo vat, the colourless leuco compound obtained on reduction being soluble in alkalis. **Aniline Black** has been claimed as a dye of this class, but there are serious objections to such being the case. Thus while **Aniline Black** is a remarkably stable compound, the indamines and indophenols are very sensitive to the action of acids and decompose by hydrolysis, giving quinones:

$$(OH) \text{ or } H_2N\langle\rangle{-}\underset{H_2}{N} \vdots \overset{=}{\underset{O}{}} \langle\rangle \overset{=}{\underset{O}{}} \vdots \underset{H_2}{NH}$$

Indamines and indophenols are readily obtained by oxidation of corresponding di-p-substituted diphenylamine derivatives. Thus p-p-diamido-diphenylamine is the leuco compound of **Phenylene Blue**, and is obtained from it on reduction.

CHAPTER XXII

AZINE, OXAZINE AND THIAZINE DYESTUFFS

THESE dyestuffs differ from each other very little in general properties. The commercial products are mainly basic dyestuffs.

They all give leuco compounds on reduction, which reoxidise in presence of air. Hence the volumetric estimation with titanous chloride is done in a CO_2

atmosphere. They have the following characteristic groupings:

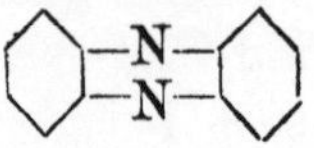

—N—
—N— **Azine**

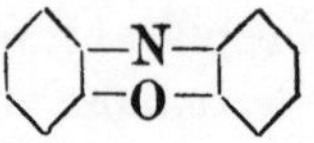

—N—
—O— **Oxazine**

—N—
—S— **Thiazine**

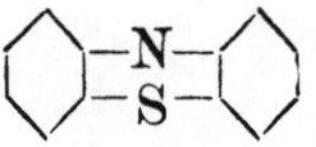

Azine Dyestuffs.—Azonium bases are derived from azines by linking an organic radicle to one of the nitrogen atoms, which is thereby changed from trivalent to pentavalent nitrogen. These bases are themselves dyestuffs, but are too feeble for commercial use without the introduction of auxochrome groups. A new branch of azines in the anthraquinone series possesses valuable vat-dyeing properties and other distinctive features which require separate treatment (see Anthracene Derivatives). The introduction of auxochrome groups into azines and azonium bases has furnished a large number of commercial dyestuffs. The first dyes of this class, safranines and indulines, were obtained by purely empirical methods. Their production by methods of oxidation involved the intermediate formation of indamines. Thus, when an ortho-amido indamine is heated in aqueous solution, it is transformed to an azine, i.e., changes from blue to red, the leuco compound formed intermediately, oxidising in the presence of air.

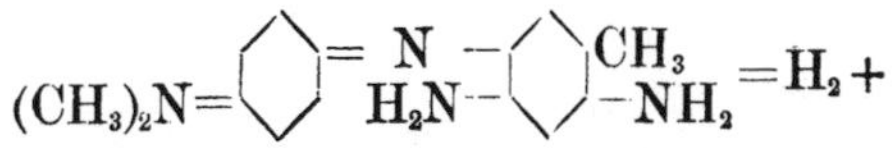

$(CH_3)_2N=C_6H_4=N-C_6H_2(CH_3)(NH_2)\cdot H_2N =H_2+$

(Toluylene Blue)

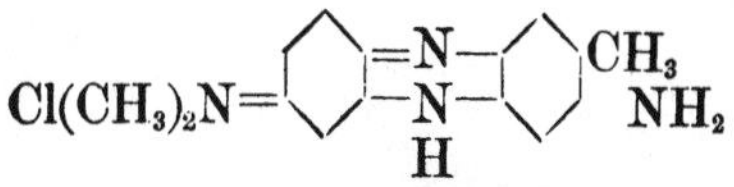

$Cl(CH_3)_2N=C_6H_3{=N- \atop -N(H)-}C_6H_2(CH_3)NH_2$

(Toluylene Red)

Neutral Red (C.)

The simplest azine type, diphenazine, obtained by heating pyrocatechol and o-phenylene diamine, may be written

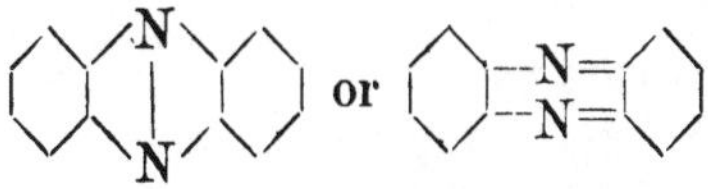

$C_6H_4{N \atop N}C_6H_4$ or $C_6H_4{-N= \atop -N=}C_6H_4$

Azines appear to be tautomeric bodies possessing the symmetrical type of constitution in the almost colourless free state. The ortho-quinonoid constitution (quinoxaline) is ascribed to the intensely coloured salts of azine bases, e.g.,

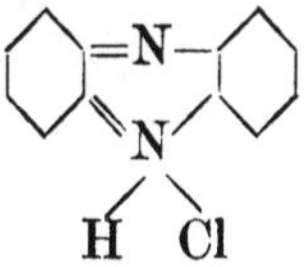

$C_6H_4{=N- \atop =N(H)(Cl)-}C_6H_4$ diphenazine hydrochloride

In many azine dyestuffs the presence of auxochrome groups permits of a p-quinonoid structure being assigned, as well as the ortho type shown above, e.g., **Neutral Red.**

These dyestuffs are, however, generally written with an ortho-quinonoid constitution

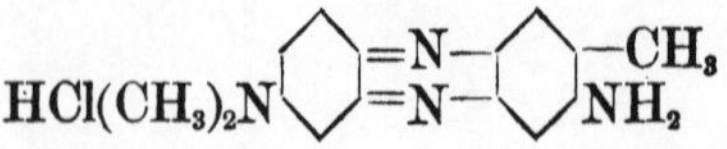

$$HCl(CH_3)_2N \langle \rangle =N- \langle \rangle -CH_3 \quad =N- \; NH_2$$

or $(CH_3)_2N$ ... $=N-$ ... CH_3, $-NH_2$, N, H Cl

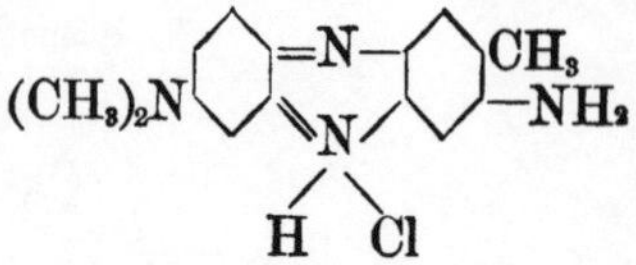

Then, again, an alternative is offered, i.e., to which nucleus the quinone bonds should be attached. It is often difficult to decide among the existing possibilities in assigning quinonoid formulae to azine dyestuffs, while the symmetrical formula usually can be safely assigned, for the time being at any rate.

The azine class can be subdivided into certain well marked groups, typical members of which will suffice.

The quinoxaline class only contains one member. The parent substance is quinoxaline or phenazine.

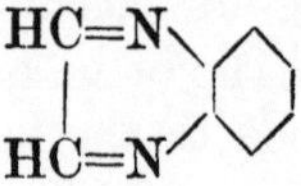

$HC{=}N$
$HC{=}N$

When phenanthraquinone is condensed with o-amido diphenylamine in glacial acetic acid solution, a basic orange yellow dye is obtained, **Flavinduline O**, II (B.).

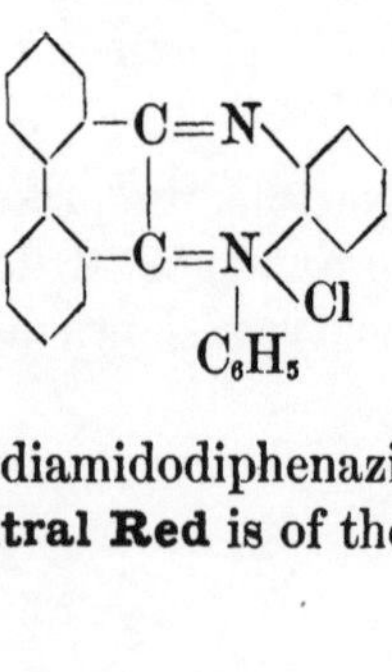

$-C{=}N$
$-C{=}N$ Cl
C_6H_5

The mono and diamidodiphenazines are called **eurhodines**, e.g., **Neutral Red** is of the eurhodine class.

The mono and di-oxy derivatives corresponding to eurhodines are called **eurhodols.**

The azonium class derived from phenylphenazonium chloride includes the **aposafranines** and **aposafranols,**

=N—
N
C_6H_5 Cl

which are respectively mono-amido derivatives of phenylphenazonium chloride and monoxy derivatives. (The auxochrome groups are meta to the pentavalent N atom.)

The **safranines** are diamido-phenylphenazonium compounds. In a typical safranine both auxochrome groups are in the m-position to the pentavalent N atom. The corresponding di-oxy derivatives are called **safranols.** Bodies of the safranine type, however, having the amido groups phenylated, are called **mauveines**, while the introduction of other phenyl-amido groups into mauveines gives **indulines.**

The aposafranines derived from naphthophenazine have also been given special names according to the position of the amido group. When this is in the naphthalene ring they are called **rosindulines**, e.g., **Induline Scarlet** (B.) is a typical rosinduline.

N
H_2N CH_3
N
Cl C_2H_5

With the amido group in the phenyl nucleus the **isorosindulines** are obtained, e.g., **Neutral Blue** (C.).

[Structural formula: naphthophenazonium ring system with N, N, $N(CH_3)_2$, Cl, C_6H_5]

These dyes are all basic dyes and find their chief use on cotton, being applied on a tannin mordant.

Induline Scarlet also finds use as a catalyst along with formaldehyde-hydrosulphite reducing agents. By its use it is possible to discharge **Naphthylamine Bordeaux** in calico printing, which is unsatisfactorily performed by use of formaldehyde-hydrosulphites alone. **Rongalite Special** (B.), and certain other "special" commercial brands of these reducing agents, contain the above dyestuff as catalyst.

The **Nigrosines** are grey and black dyestuffs, of which little is known as to constitution, except that they are related to the indulines.

Eurhodines.—This small and unimportant class may be shortly dealt with. The method of obtaining **Neutral Red** above described is typical. Any of the alternative methods of synthesis available for the required indamine may be employed.

Further condensation proceeds readily on heating in aqueous solution as in the case of **Neutral Red**.

The eurhodines are weak bases forming red mono-acid salts and green di-acid salts. Only the former come into consideration for dyestuffs as the latter are dissociated in aqueous solution.

The eurhodols obtained by replacement of amido

groups by oxy groups in eurhodines have no commercial importance.

Safranines.—The elucidation of the constitution of these dyestuffs presented great difficulty, and is mainly due to Hofmann, Witt, Nietzki and Bernthsen. The following facts are to be considered in adopting a formula. Indamines (blue) are well marked intermediate stages in safranine (red) formation. Safranines are obtained by heating indamines with primary monamines; also by oxidation of p-p′-diamido-diphenylamine in presence of primary amines, by oxidation of a p-diamine in presence of primary amines, by oxidation of a p-diamine along with a derivative of m-amidodiphenylamine, and by the commercial method of oxidising a molecule of a p-diamine with two molecules of a monamine.

By this last method it is found that the following conditions are essential to obtain a safranine: (i) the p-diamine employed must have one amido group unsubstituted, one molecule of this body being required; (ii) two molecules of the same or different monamines are required, of which one must have the p-position unsubstituted, while in the other it may or may not be substituted, but the amido group itself must be unsubstituted, i.e., a primary amine.

Witt proposed an asymmetrical formula for safranine:

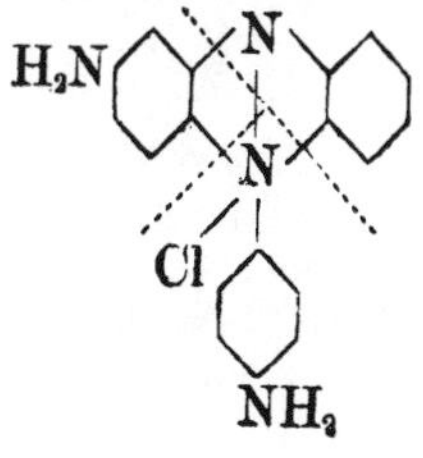

H_2N N N Cl NH_2

The dotted cleavage lines show how well it agrees with the main facts of safranine synthesis as laid down above. It has, however, been dropped and the formula proposed by Bernthsen is generally adopted:

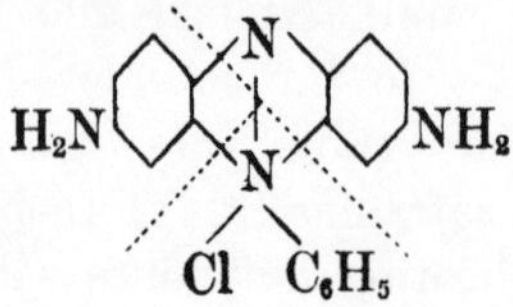

N
H_2N NH_2
N
Cl C_6H_5

It was found that the number of substituted amido derivatives according to the asymmetrical formula could not be prepared. Also m-amido-o′. o′-dinitro-diphenylamine could be condensed with p-phenylene diamine to give a safranine, which is not in accordance with asymmetric substitution. Again, the same phenylsafranine is obtained in two ways: (i) by oxidation of m-amido-diphenylamine along with p-amidodiphenylamine, (ii) by oxidation of m-phenyl-amidodiphenylamine and p-phenylene diamine. These facts are not explained by the asymmetric[1] formula but quite agree with the Bernthsen formula. This latter is adapted to express either ortho or para quinone structure. Thus the simple safranine written above may be expressed:

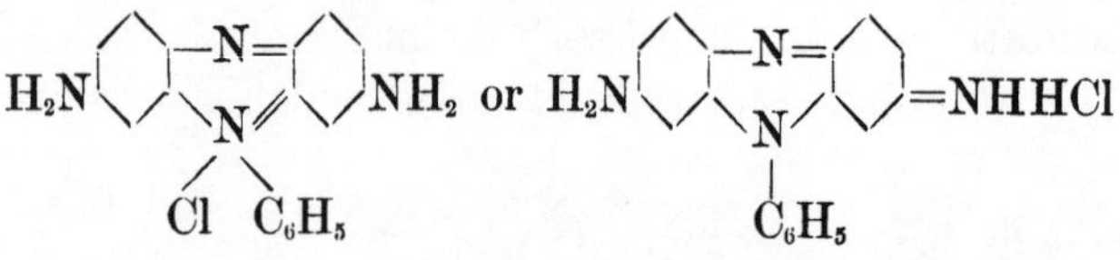

—N=
H_2N NH_2 or H_2N =NHHCl
N N
Cl C_6H_5 C_6H_5

[1] The asymmetric formula has recently been revived by Barbier and Sisley who stated that the commercial dye is mainly of this type. Their claim however has since been refuted by Hewitt and his collaborateurs who found their oxyaposafranone to be identical with the body obtained from p-nitrosophenol and m-oxydiphenylamine.

The main trend of evidence is in favour of the o-quinone formula. Safranine is strongly basic, and the base itself is not obtained in the usual way, namely, by alkali treatment of the salt. It may however be obtained by oxidation with silver oxide of the following diphenylamine derivative:

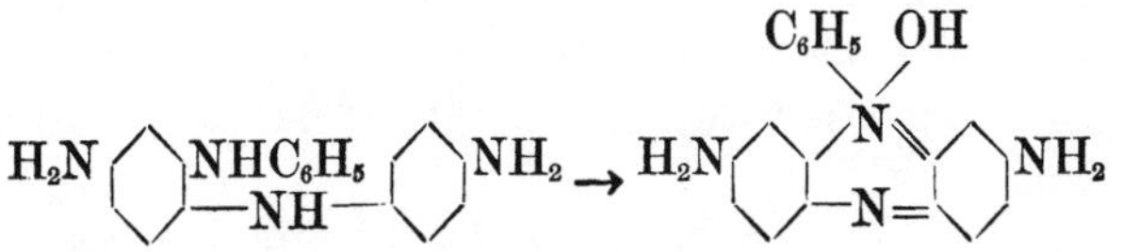

C_6H_5 OH

N

H_2N NHC_6H_5 NH_2 → H_2N NH_2

—NH— —N=

It is strongly basic, due to the pentavalent nitrogen atom which is an essential part of the ortho quinonoid structure.

The ready reoxidation of the leuco compound of safranine in air agrees also with the general behaviour of o-quinone dyestuffs in this respect. According to the p-quinone formula safranine is an imide. It is actually the case that by ordinary methods of diazotisation only one amido group can be diazotised. When this is done and the diazo body boiled with alcohol aposafranine is obtained which cannot be further diazotised by the general methods.

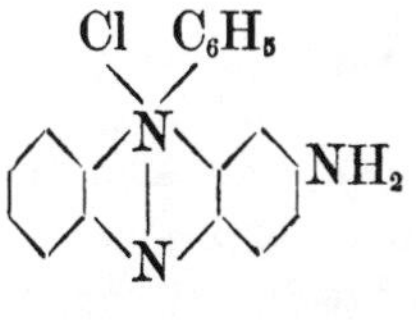

Cl C_6H_5

N

NH_2

N

However, by treatment in strong acid solution (green) this body can be diazotised, and also safranine can be diazotised in both amido groups, and on boiling with alcohol, phenylphenazonium chloride is obtained.

Cl C_6H_5
N
—N=

This body is reconverted into safranine base by treatment with ammonia.

Safranine base may also be obtained by treatment of the sulphate with baryta. It crystallises in green leaflets.

It may further be stated that safranine base on heating can be obtained almost free from oxygen, which is in accordance with the p-quinonoid formula.

C_6H_5
H_2N N =NH
—N=

On the other hand certain chlorine substitution products of safranol clearly support the ortho quinone structure.

C_6H_5 Cl C_6H_5
O N OH HO N OH
—N= —N=

safranol base (or safranone) hydrochloride

The formula for aposafranol may be written from what has been explained above.

The formation of safranine by the oxidation method of synthesis is apparently due to formation of quinone derivatives, by oxidation of the p-diamine used, which then form diphenylamine derivatives by addition of molecules of amines. These diphenylamine derivatives by oxidation give indamines which have again a quinone structure and allow of further addition, giving phenyl-amidodiphenylamines of the type shown on page 229 from which safranine base is obtained. Further oxidation in presence of acid gives the dyestuffs found in commerce. The whole process bears considerable resemblance to the synthesis of Magenta in respect of the part played by quinone structure. (See p. 191.)

The constitutions of other dyestuffs of the azonium class have been assigned on similar evidence to that discussed above for safranine.

Commercial **Safranine** (many brands) is obtained by oxidation of equal molecules of p-tolylene diamine and o-toluidine (obtained mixed by reduction of amido-azotoluene), to give an indamine. The required extra molecule of monamine is then added, i.e., aniline or o-toluidine. The recovered oils from the Magenta melt are rich in o-toluidine, and may be used for Safranine manufacture. Potassium bichromate in hot acid solution is the oxidising agent. Mauve and other azines are also formed. Chalk is added to precipitate impurities and the dyestuff salted out of the filtrate. The salt of the dyestuff formed with hydrochloric acid is crystalline.

$Cl\quad C_6H_5$

N

$H_2N \qquad\qquad NH_2$

$H_3C \quad -N= \quad CH_3$

and

$Cl\quad C_6H_4 . CH_3$

N

$H_2N \qquad\qquad NH_2$

$H_3C \quad -N= \quad CH_3$

Methylene Violet BN, etc. (M.) is the asymmetric (—N($CH_3)_2$) dimethyl-phenyl safranine. Other dyes of this class are **Tannin Heliotrope** (C.), **Rhoduline Violet** (By.), **Amethyst Violet** (K.), etc.

Of the phenyl (C_6H_5 . NH—) safranines or mauveines few are important.

Mauve itself possesses historical interest, being the first of the coal tar dyestuffs. It was obtained by Perkin in 1856 by oxidation of a crude aniline containing mixed toluidines.

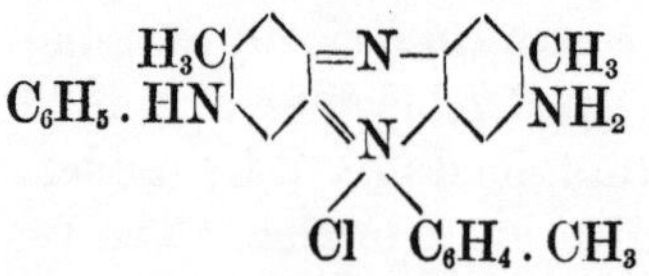

H_3C =N— CH_3
C_6H_5 . HN N NH_2
Cl C_6H_4 . CH_3

Mauve or Rosolane

It still finds some small use as a basic dye, mainly on account of its comparatively good fastness to light, which exceeds that of **Methyl Violet**, on that account it finds some use for blueing or "white-dyeing" of silk.

Indazine M, etc. (C.) is a blue basic dye of the mauve class, as is also **Metaphenylene Blue**.

Milling Blue (K.) is a sulphonated di-naphthosafranine or rosinduline of the mauve type.

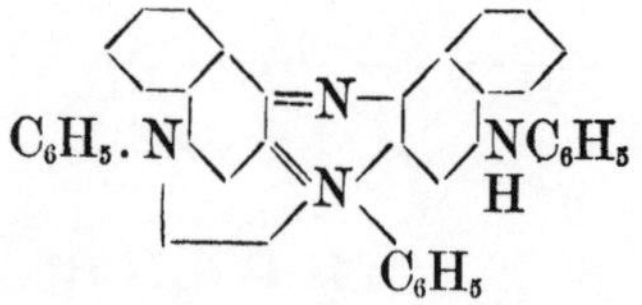

=N—
C_6H_5 . N N NC_6H_5
H
C_6H_5

Magdala Red is a basic dyestuff of the safranine class.

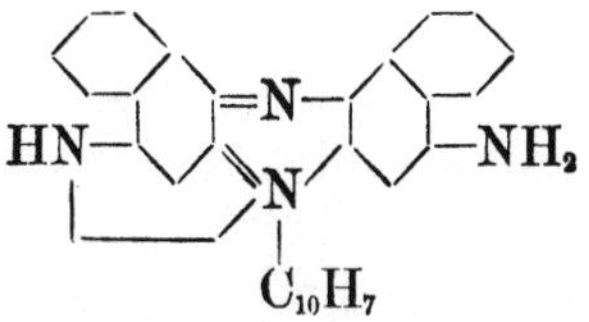

$$HN{-} \quad {=}N{-} \quad {-}NH_2$$
$$N$$
$$C_{10}H_7$$

It is the naphthyl homologue of phenyl safranine. It exhibits fluorescence, and finds a small use on silk.

Certain rosindulines (aposafranine class) are also important, and categorically should have been dealt with before the safranines. A good deal of evidence, analogous to that brought up in discussing the constitution of safranines, has been accumulated in favour of a similar type of formula for rosindulines. The simplest rosinduline is obtainable by condensing 4-amido-β-naphthoquinone with o-amido-diphenylamine, whereby two molecules of water are split off, giving the base.

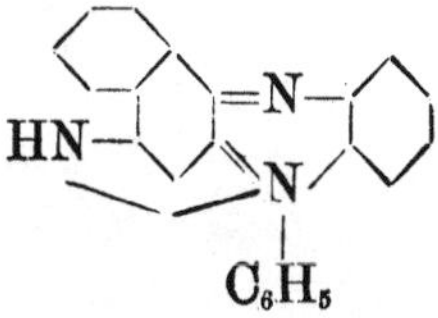

$$HN{-} \quad {=}N{-}$$
$$N$$
$$C_6H_5$$

The rosindulines are sometimes obtained commercially by heating simple azo compounds with α-naphthylamine hydrochloride or aryl-azo-α-naphthylamines with aniline and aniline hydrochloride (i.e., a primary amine), e.g., **Induline Scarlet** already mentioned is obtained from the azo compound of monoethyl-p-toluidine by melting with α-naphthylamine hydrochloride. Anilido-naphthoquinones and aniline-α-naphthylamine are also used to give rosindulines by similar heating methods. The process involves the formation of intermediate bodies analogous to those found in the induline melt (see later). These have been isolated by stopping the melt before completion of the reaction. Thus. by melting benzol azo-α-naphthylamine with aniline and aniline hydrochloride, phenylrosinduline is obtained finally, by intermediate formation of a trianilido quinone:

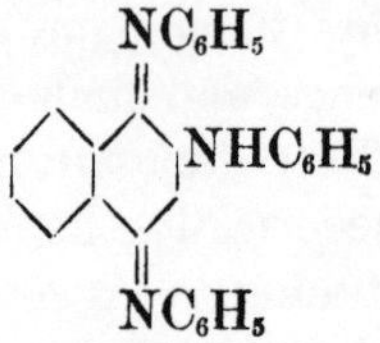

NC_6H_5

NHC_6H_5

NC_6H_5

This body can be easily prepared in good yield from β-naphthoquinone-4-sulphonic acid by heating with aniline. The first stage is reached in aqueous solution, but the second only by melting.

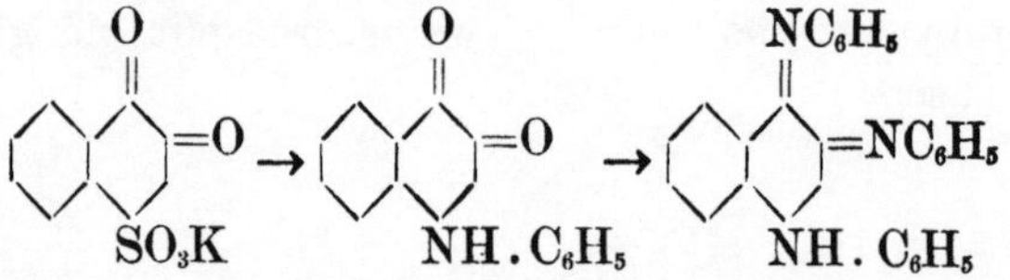

O ... =O ... SO_3K → O ... =O ... $NH.C_6H_5$ → NC_6H_5 ... $=NC_6H_5$ ... $NH.C_6H_5$

This trianilido quinone is a tautomeric form of the intermediate product above written. By further melting with aniline and aniline hydrochloride up to 160° C. phenylrosinduline is obtained. Thus the trianilido quinone

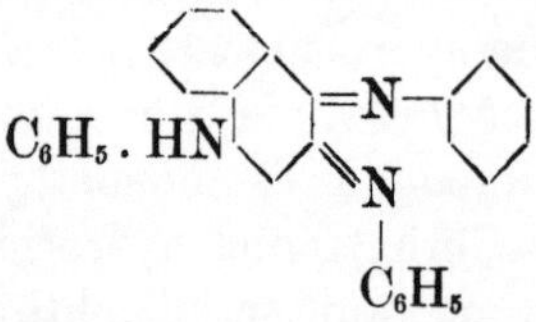

$C_6H_5.HN$ ... =N– ... N ... C_6H_5

by direct condensation gives leuco-phenylrosinduline, which readily oxidises in presence of air to the dye-stuff base.

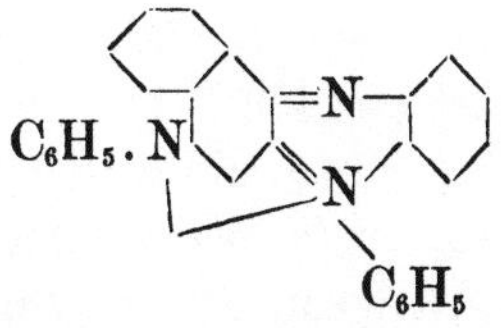

=N–
$C_6H_5.N$
N
C_6H_5

Rosinduline base.

The excess of aniline is dissolved out of the melt with the required amount of hydrochloric acid and phenylrosinduline filtered off. This product finds commercial use as the disulphonic acid **Azocarmine G in paste** (B.), which is an acid dyestuff difficultly soluble in water. The monosulphonic acid is first prepared by heating up to 95° C. with oil of vitriol. It is then purified by treatment with caustic soda and washing, and further sulphonated. This intermediate purification results in a better shade in the final product.

Phenylrosinduline hydrochloride is readily dissociated hydrolytically, and is not used as a basic dye.

Azocarmine B (B.) is the trisulphonic acid of phenylrosinduline.

By heating this dyestuff with water at 160° to 180° C. the monosulphonic acid of a rosindone is obtained.

Rosinduline 2G (K.) is the monosulphonic acid of a rosindone.

=N–
O
N
C_6H_5

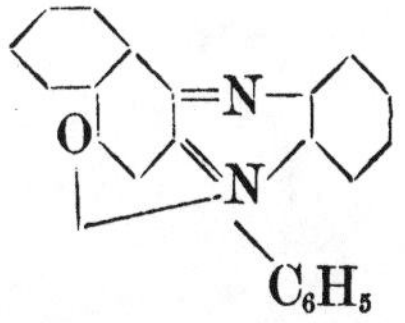

Rosinduline G (K.) is a very similar body. These are acid dyestuffs for wool.

Neutral Blue (C.) already mentioned as an isorosinduline is obtained by the action of nitroso-dimethylaniline hydrochloride on phenyl-β-naphthylamine.

Indulines.—This class includes blue, violet and blue-black dyestuffs obtained by heating amidoazobenzene with aniline and aniline hydrochloride to the boiling point, or under pressure. Almost any azo or nitroso compound on treatment in this way will give indulines. If p-phenylene diamine be added, basic indulines are obtained. The usual products are insoluble in water but soluble in alcohol, and on sulphonation give water-soluble acid dyes.

It has been found that the formation of quinone anilides plays an intermediate rôle in induline synthesis. Thus azophenine has been isolated and identified by stopping the melt and extracting with alcohol. It forms dark red crystals of formula

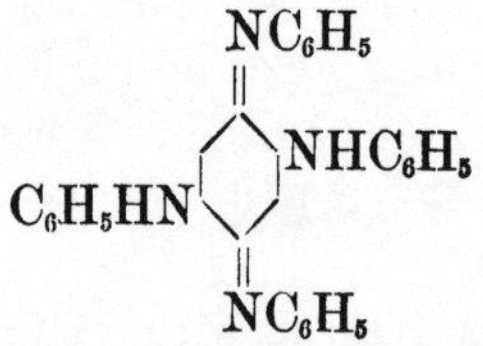

NC_6H_5
‖
C_6H_5HN — ring — NHC_6H_5
‖
NC_6H_5

The changes involved in the formation of this body from amidoazobenzene are probably of the following type:

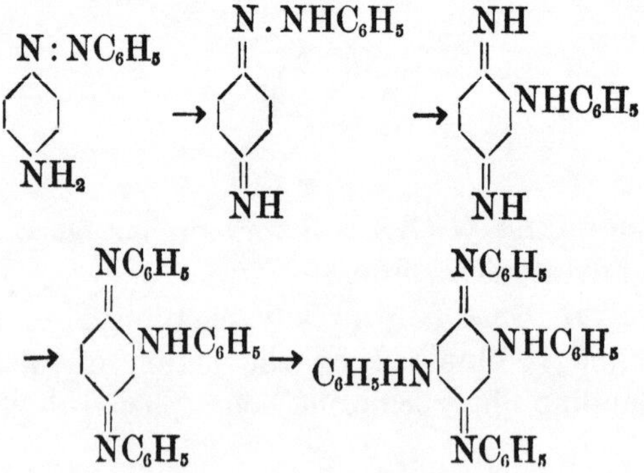

$N:NC_6H_5$ / ring / NH_2 → $N.NHC_6H_5$ ‖ ring ‖ NH → NH ‖ ring (NHC_6H_5) ‖ NH

→ NC_6H_5 ‖ ring (NHC_6H_5) ‖ NC_6H_5 → NC_6H_5 ‖ C_6H_5HN ring (NHC_6H_5) ‖ NC_6H_5

The removal of two H atoms as required for the final stage of this process, appears to be accomplished by reduction of a certain amount of amidoazobenzene, which would account for the presence of some p-phenylenediamine found in the melt.

Quinone anilides of the azophenine type readily condense further on raising the temperature from 100° to 130° C. thus:

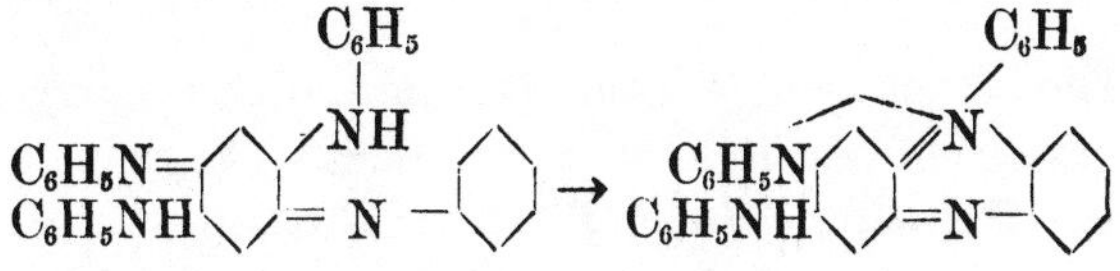

$$\begin{matrix} & C_6H_5 & & & & C_6H_5 \\ C_6H_5N= & NH & & \rightarrow & C_6H_5N & N \\ C_6H_5NH & =N- & & & C_6H_5NH & =N- \end{matrix}$$

The process of induline formation should be compared with that of rosinduline synthesis already described.

Further heating results in the introduction of more anilido groups.

The following are well known members of the induline class obtained by more or less modified processes, similar to the one described:

Indamine Blue (M.). A basic dye used instead of indigo.

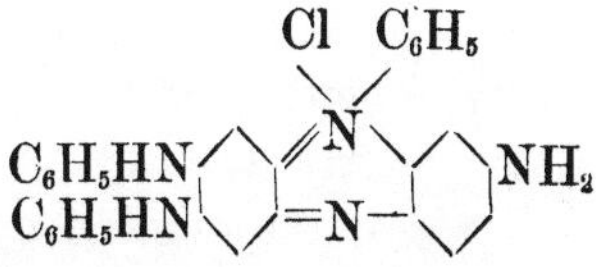

$$\begin{matrix} & Cl \quad C_6H_5 & \\ C_6H_5HN & N & NH_2 \\ C_6H_5HN & =N- & \end{matrix}$$

Other **Indamine Blue** brands are obtained by heating nitroso-dimethylaniline with o- or p-toluidine.

Induline B, 3B, etc. (many brands). The spirit-soluble brands are hydrochlorides of the same type as **Indamine Blue** above and of the following bases:

$$C_6H_5$$
$$C_6H_5N \quad N \quad NHC_6H_5$$
$$C_6H_5HN \quad =N- $$

$$C_6H_5$$
$$C_6H_5N \quad N \quad NHC_6H_5$$
$$C_6H_5HN \quad =N- \quad NHC_6H_5$$

The water-soluble acid dyes are obtained by sulphonation of spirit-soluble indulines. Various **Fast Blues**, **Printing Blues**, **Acetine Blues**, etc., are of this type. Besides being acid dyestuffs suitable for wool and silk, they possess residual basic properties allowing of fixation on tannin-mordanted cotton.

Nigrosines are obtained by heating nitrobenzene with aniline and aniline hydrochloride in presence of metallic iron (filings) to 180° C. Nitrosophenol also gives dyes of this class when heated with aniline and aniline salt. They are used for pigments, shoe polishes, etc. The water-soluble acid dyes of this class are obtained by sulphonation of the blue-black or black nigrosine bases.

Aniline Black. In 1834 Runge obtained insoluble black compounds by oxidation of aniline with chromic acid. A similar precipitate was obtained by Perkin in making Mauve. It was not applied in dyeing till 1862, when Calvert obtained a fast black on cotton by oxidation of aniline in contact with the cotton fibre. Aniline black is largely produced in this way, and itself is not a commercial dyestuff.

There are three chief methods in use: (1) the so-called single-bath black is used for cotton hanks, which are treated with a cold mixed solution of aniline, potassium bichromate and sulphuric acid. Later the temperature is gradually raised. A large amount of black is precipitated in the bath and a considerable quantity of chrome is deposited on the fibre as well as aniline black itself. (2) The so-called "aged" or oxidation black is applied to cotton by padding with a solution containing aniline, aniline salt, potassium chlorate and a copper salt, the latter acting as oxygen carrier. The dried material is then "aged" in a warm moist chamber and becomes green, due to formation of emeraldine. It is given a bath of bichromate to complete the oxidation of emeraldine to aniline black. Attempts have been made to overcome the tendering of cotton in this process by replacement of the chlorate by air oxidation in presence of a trace of a catalyser, e.g., p-phenylene diamine. (3) The steam black can be reserved, and is used in calico printing. The padding liquor contains aniline salt, potassium chlorate and ferrocyanide, and by printing on a thickened alkaline paste after drying the development of aniline black is prevented from taking place under the reserve paste in the subsequent steaming.

All these processes show similar stages in aniline black formation, namely, first a green compound called emeraldine, then a blue-black body turned green by sulphur dioxide or mineral acids. This base has been called nigraniline or greenable black. Finally, ungreenable aniline black or pernigraniline is obtained, which is not affected by sulphur dioxide or boiling with dilute mineral acids, and is a very stable body. On reduction it gives an almost colourless leuco body which re-oxidises in air, and on oxidation a very good yield of p-benzoquinone. Complete reduction or distillation with zinc dust gives para diamines and amido-diphenylamines. The recent researches of Willstaetter and Green and their collaborateurs have increased the knowledge of aniline black. The first stage in the oxidation of aniline is the formation of phenyl quinone di-imide

$$C_6H_5{-}N{=}\langle\quad\rangle{=}NH$$

a yellow body which polymerises in presence of acids to give emeraldine $C_{24}H_{20}N_4$. Further oxidation of this base gives a red compound which it is claimed by Willstaetter can polymerise to give finally aniline black. Green thinks Willstaetter's products are not those obtained in the commercial production of aniline black. The lengthy indamine formula deduced from polymerisation would not possess the stable properties of aniline black. Finally, analyses of aniline black vary, no doubt due to the black product consisting of mixtures of nigraniline with pernigraniline, some free aniline is necessary to get the fully formed black by oxidation of nigraniline, according to Green.

It has long been recognised that aniline black shows great similarity to the azines both in method of formation and properties. A considerable difficulty, however, has been experienced in the fact that the degradation products obtained from aniline black contain no ortho derivatives. The trend of evidence is in the main towards an azine structure for aniline black, to which Green has ascribed the following formula:

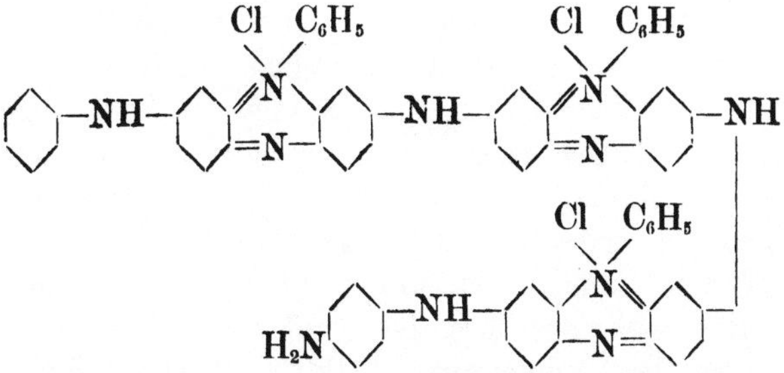

Cl C_6H_5 Cl C_6H_5

—NH— N —NH— N —NH

=N— =N—

Cl C_6H_5

—NH— N

H_2N —N=

See *Jour. Soc. Dyers*, contributions by A. G. Green and others, pp. 105 and 338, 1913. Also *Aniline Black*, Noelting and Lehner.

Diphenyl Black Base P (M.) consists of amido-diphenylamine and on oxidation gives an aniline black. **Diphenyl Black Oil DO** (M.) is a mixture of the base and aniline. The use of these bodies demands less strong oxidation on the fibre, and therefore reduces the risks of tendering the fabric.

Paramine Brown is a chocolate brown produced on the fibre from p-phenylene diamine in a similar way to the production of aniline black. The diamine is sold for fur dyeing, etc., as **Ursol D** (Ber.), **Furreïn D** (J.), etc. Other **Ursol** and **Furreïn** dyeings are produced by p- or m-amidophenol. The **Furrole** (C.) and **Nako** (M.) colours belong to the same class.

Oxazine Dyestuffs.—These dyestuffs are derivatives of diphenoxazine, which is obtained by condensing o-diamidophenol with pyrocatechin.

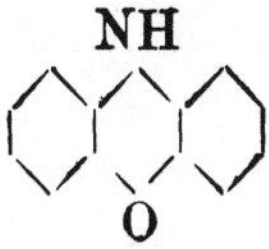

NH

O

By introducing auxochrome groups into this body leuco oxazine dyestuffs are obtained which oxidise in air to give the dyestuffs. The class exhibits great similarity to the azines in most respects. They are prepared from o-hydroxy derivatives of indamines or indophenols, or with these bodies as intermediate stages involved in their production. The various methods of indamine synthesis may be employed to reach this stage, further condensation taking place readily on heating, by quinone change; e.g., nitrosodimethylaniline and resorcin give

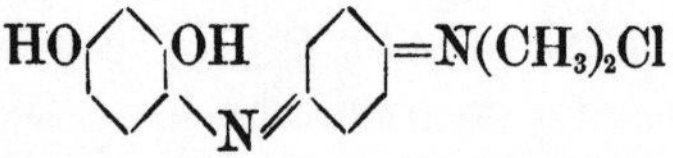

$$HO \quad OH \qquad =N(CH_3)_2Cl$$
$$N$$

This is produced on the cotton fibre by printing with a paste containing resorcin and the nitroso body with tannin as mordant. The cloth is dried and steamed; the indamine which is first formed condenses further to give **Nitroso Blue** or **Resorcine Blue**:

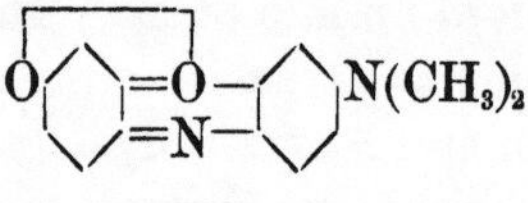

$$O \quad =O- \quad N(CH_3)_2$$
$$=N-$$

It will be noted that the adoption of an o-quinonoid formula for oxazine dyestuffs demands the assumption of tetravalent oxygen.

The first commercial dyestuff of this class was **Naphthol Blue D**, **R**, **New Blue R** or **Meldola's Blue** discovered in 1879 by Meldola. It is obtained by adding gradually nitrosodimethylaniline hydrochloride to β-naphthol dissolved in its own weight of glacial acetic acid, or three times its weight of alcohol, the temperature being maintained at 110° C. under reflux condenser. The reaction is violent at first. The pure dyestuff may be separated with alcohol, but it is usually crystallised out as the double chloride with zinc. It is used for cheap indigo blue shades on tannin-mordanted cotton.

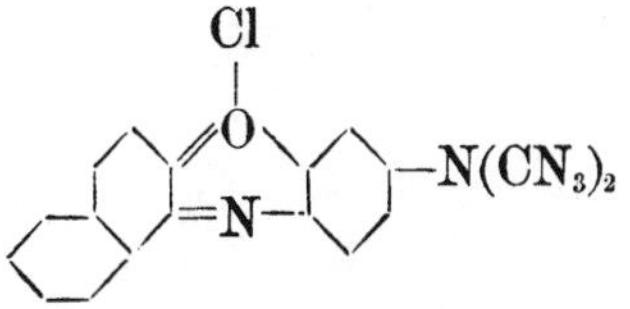

Cl
O
=N— —$N(CN_3)_2$

By further action of excess of nitrosodimethylaniline **New Blue B** is obtained.

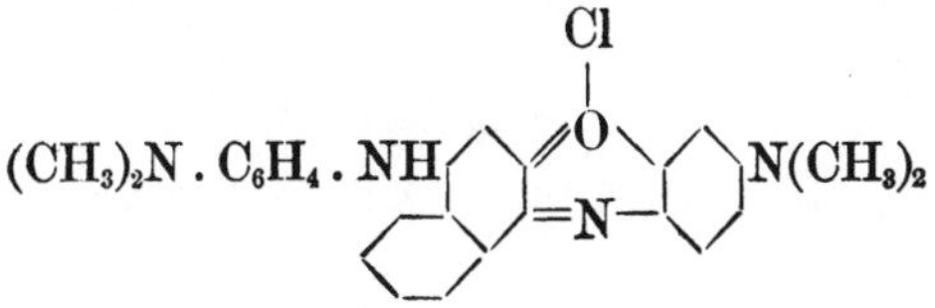

Cl
O
$(CH_3)_2N \, . \, C_6H_4 \, . \, NH$ $N(CH_3)_2$
=N—

If, instead of β-naphthol, in making **Meldola's Blue**, 2.7-dioxynaphthalene be used, **Muscarin** is obtained.

Other basic dyestuffs of this class are obtained similarly thus:

Capri Blue GON (By.), nitrosodimethylaniline and dimethyl m-amidocresol.

New Methylene Blue GC (C.), dimethylamine and **Meldola's Blue**, and subsequent oxidation.

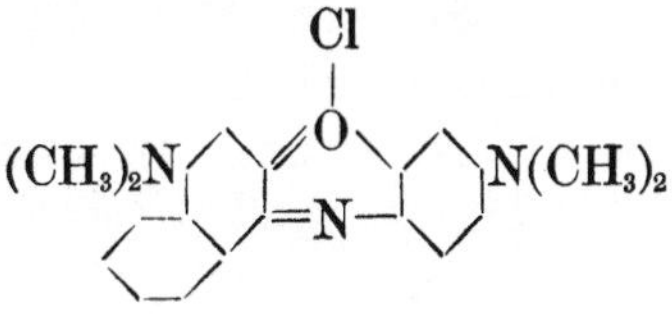

Cl
O
$(CH_3)_2N$ $N(CH_3)_2$
=N—

Nile Blue A (B.), nitrosodiethyl-m-amidophenol and α-naphthylamine.

$$SO_3H$$
$$H_2N \quad O \quad N(C_2H_5)_2$$
$$=N-$$

Using benzyl-α-naphthylamine in the above condensation, one gets **Nile Blue 2B** (B.).

Fast Black (L.) is obtained by action of nitrosodimethylaniline on m-oxydiphenylamine.

Beside the above basic dyestuffs some important mordant dyes belong to this group.

By acting on gallic acid in boiling methyl alcohol solution with nitrosodimethylaniline hydrochloride, **Gallocyanine** (many brands) is obtained.

$$OH$$
$$(CH_3)_2N \quad -O= \quad O$$
$$-N=$$
$$COOH$$

It gives violet blue chrome lakes, and is extensively used on wool and in calico printing. Certain brands (powder) are bisulphite compounds. On heating with aniline, followed by sulphonation, it gives **Delphin Blue B** (By.), an acid mordant dyestuff, the —COOH group being replaced by $—NH—C_6H_4—SO_3NH_4$.

Modern Blue, **Modern Cyanine**, **Indalizarin**, **Prune**, etc., are **Gallocyanine** derivatives.

Prune pure (S.) is obtained by using, instead of gallic acid (—COOH), the methyl ether ($—COOCH_3$).

Gallamine Blue paste (By., G.) is obtained by

using gallamide (—CO NH_2) instead of gallic acid (COOH).

Gallanil Violet (D.H.) is obtained by use of the anilide of gallic acid (—CO . NH . C_6H_5).

Corein is the diethyl analogue of **Gallamine Blue.**

There are many others of this class.

The oxy-derivatives of the oxazine class corresponding to safranones are of no commercial value.

Thiazine Dyestuffs.—These are closely allied to the azine and oxazine groups. They contain one atom of sulphur, which is regarded as tetravalent in the o-quinonoid formula. **Lauth's violet** was the earliest known colouring matter of this class. It was obtained in 1876 by oxidation of p-phenylene diamine with ferric chloride in acid solution in presence of sulphuretted hydrogen. The violet thus obtained has no commercial value, but substitution of dimethyl-p-phenylene diamine for the simple diamine by Caro gave **Methylene Blue.** The constitution of the thiazines has been largely built up from the work of Bernthsen on these two dyestuffs. He nitrated thiodiphenylamine, obtained by heating diphenylamine with sulphur, and on reduction obtained the leuco base of Lauth's violet which oxidised in presence of air to give the dyestuff.

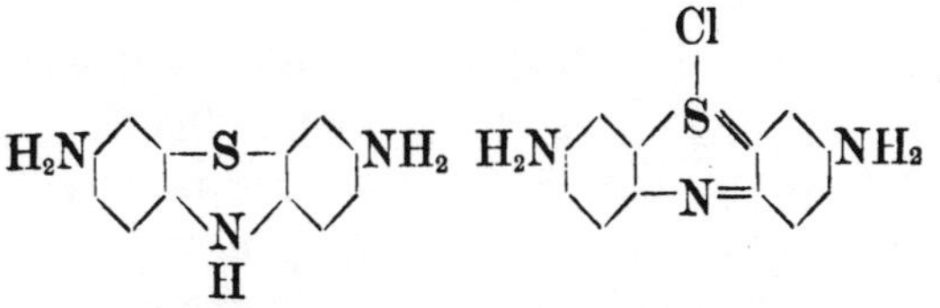

H_2N—S—NH_2 (N—H) H_2N—S(Cl)—N=—NH_2

leuco base   Lauth's Violet

Hence also the constitution of **Methylene Blue**:

$$(CH_3)_2N-C_6H_3<\genfrac{}{}{0pt}{}{S(Cl)=}{-N=}>C_6H_3-N(CH_3)_2$$

Bernthsen devised a new and cheaper synthesis of this dyestuff. Nitrosodimethylaniline is reduced with iron filings, and the resulting solution of dimethyl-p-phenylene diamine is oxidised in dilute acid solution with bichromate, in presence of sodium thiosulphate. A thiosulphonic acid is formed:

$$C_6H_4(N(CH_3)_2)(NH_2) + HS.SO_3H + O = C_6H_3(N(CH_3)_2)(NH_2)S.SO_3H + H_2O$$

This body can be obtained as white crystals, but is never isolated from solution in the manufacture of **Methylene Blue.** An extra molecule of dimethylaniline is added, and with further oxidation gives an indamine thiosulphonic acid.

$$(CH_3)_2N-C_6H_3<\genfrac{}{}{0pt}{}{S.SO_3}{-N=}C_6H_4=N(CH_3)_2$$

This greenish blue body on boiling with zinc chloride solution gives the dyestuff, by first losing H_2SO_4, and the leuco-azine thus formed oxidises rapidly in air, or in presence of the acid oxidising agent. The dyestuff is put out commercially as a double chloride with $ZnCl_2$.

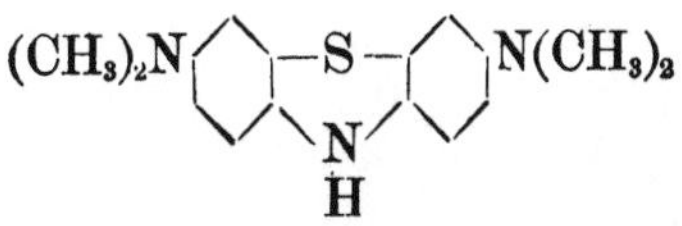

$(CH_3)_2N$ —S— $N(CH_3)_2$
N
H

leuco compound

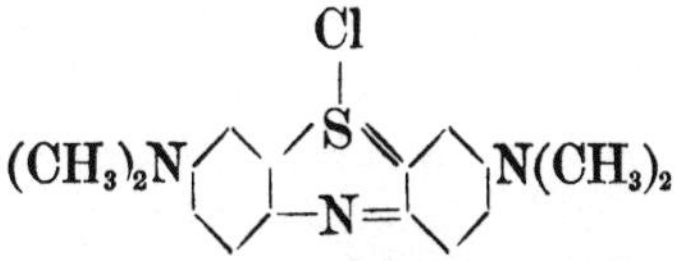

Cl
S
$(CH_3)_2N$ —N= $N(CH_3)_2$

Methylene Blue

Methylene Blue is largely used on tannin-mordanted cotton and as a basic "topping" colour, i.e., for brightening up other shades. It is faster to light than most basic dyes. It is used as a stain in microscopy, and as free base finds internal use as a medicine.

New Methylene Blue N, **GB**, **R**, etc. (C.) is obtained when mono-ethyl-o-toluidine is used as in the synthesis last described.

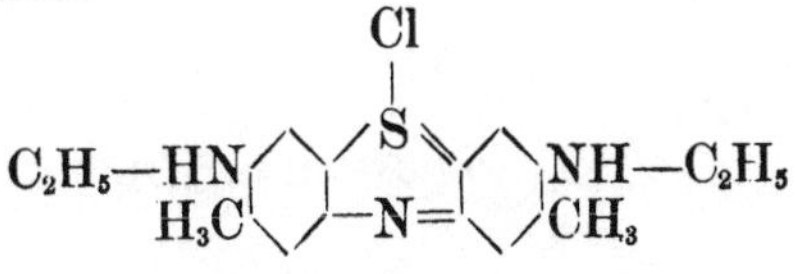

Cl
S
C_2H_5—HN NH—C_2H_5
H_3C —N= CH_3

Certain brands of this name are oxazines, see p. 243.

Methylene Green (many brands) is obtained by nitration of **Methylene Blue** with sodium nitrite and hydrochloric acid.

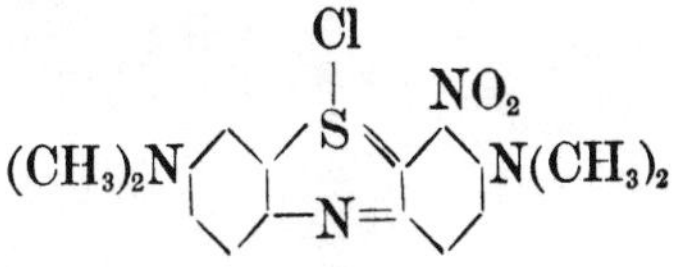

Cl
NO_2
S
$(CH_3)_2N$ —N= $N(CH_3)_2$

It is not known for certain to which nucleus the o-quinonoid bonds should be attached.

Thionine Blue O, B, etc. (M., By.) is the trimethyl-ethyl product analogous to **Methylene Blue** (a tetra-methyl product) and is similarly prepared.

The above are all basic dyestuffs. The only acid dyestuff of outstanding interest belonging to this class is **Thiocarmine R** (C.). This dyestuff is made by a thiosulphonic acid process, similar to that described for **Methylene Blue**, using p-amidoethylbenzylaniline sulphonic acid as starting point.

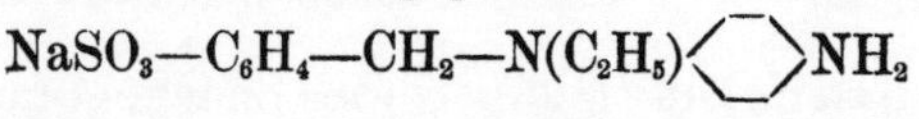

$$NaSO_3—C_6H_4—CH_2—N(C_2H_5)\langle\ \ \rangle NH_2$$

The thiosulphonic acid of this body is oxidised to form an indamine with ethylbenzylaniline sulphonic acid, which gives finally a blue acid dye:

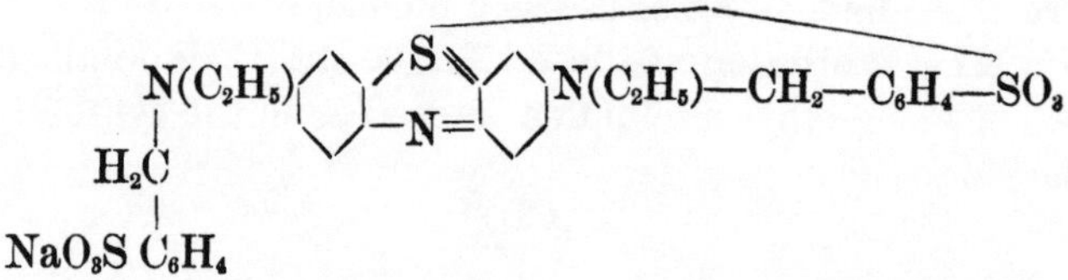

$N(C_2H_5)$ ... S ... $—N=$ ... $N(C_2H_5)—CH_2—C_6H_4—SO_3$

H_2C

$NaO_3S\ C_6H_4$

There are several important acid-mordant dyes of the thiazine class.

Indochromogen S (S.) is obtained by condensation in dilute alkaline solution of p-amidodiethylaniline thiosulphonic acid with 1.2-naphthoquinone-4.6-di-sulphonic acid.

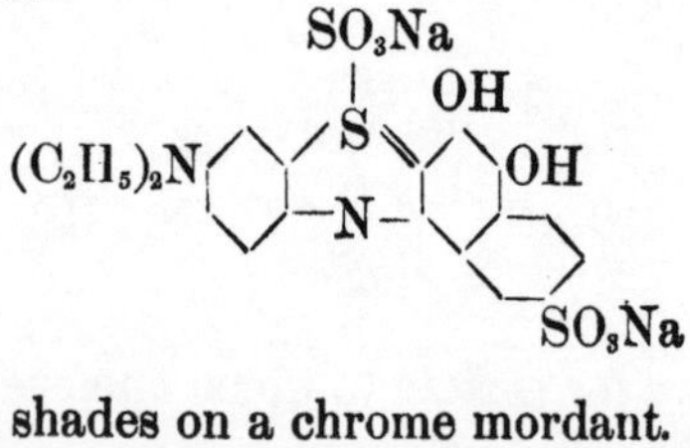

SO_3Na

OH

$(C_2H_5)_2N$ ... S ... OH

$—N—$

SO_3Na

It gives blue shades on a chrome mordant.

Brilliant Alizarin Blue G, R, etc. (By.) is obtained

by a similar method and is the dimethyl body corresponding to **Indochromogen S** (diethyl). Sulphobenzyl ethyl-p-phenylene diamine and analogous compounds give similar dyestuffs.

Additional information may be obtained from sections of Thorpe's *Dictionary of Applied Chemistry*, where may also be found patent references, etc.

CHAPTER XXIII

OXYKETONE DYESTUFFS

THE dyestuffs of this series belong to the mordant class and are mainly used in conjunction with a metallic mordant.

One of the simplest dyes of this series is **Alizarin Yellow C**, which is gallacetophenone. It is prepared by heating acetic acid with pyrogallol in presence of zinc chloride.

```
    /\—COCH3
    | |
 HO \/ OH
    OH
```

Alizarin Yellow A is the corresponding trioxy benzophenone. This dyes redder shades than the above compound.

Dyestuffs are obtained by careful oxidation of oxybenzoic acids. **Galloflavin W** (B.) is obtained by oxidation of gallic acid in alcoholic potash solution with air, followed by treatment of the product with acid.

The simplest of the diketo-dyes of this series is

Alizarin Black WR, SW, SWR, etc. (B.) is 1.2-dioxy-5.8-naphthoquinone, and is prepared from crude dinitronaphthalene (1.5- and 1.8-) by heating with a solution of sulphur sesquioxide in sulphuric acid solution, followed by hydrolysis of the 1.2-amino-oxy-α-naphthoquinone-imine by boiling with dilute acid. The insoluble **Naphthazarin** is converted into a water-soluble product by addition of sodium bisulphite.

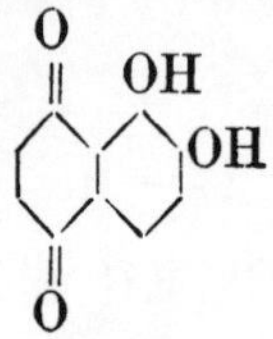

O
OH
OH
O

One of the oldest and best known members of this series is **Alizarin** which is present in madder root, which contains from two to four per cent., as a glucoside. It is a good example of those colouring matters which dye only with the aid of a mordant, and which yield various colours with different mordants (polygenetic dyes). The use of a mordant is obligatory, alizarin itself having no affinity for vegetable fibres, and only imparting a fugitive orange brown colour to animal fibres. Its compound with alumina is red, with stannous oxide orange, with chromic oxide brownish violet, with ferrous oxide blackish violet, with ferric oxide brownish black. These colours when produced on textile fibres are fast to light, washing, and milling, etc., the fastness however varying somewhat with the mordant used.

In 1869 Perkin, and also Graebe and Liebermann, produced alizarin from anthraquinone by sulphonation with fuming sulphuric acid, followed by caustic fusion in presence of an oxidising agent. In the manufacture of alizarin it is only necessary to obtain anthraquinone-β-sulphonic acid, since this compound on heating with caustic soda gives alizarin.

$$C_6H_4\begin{matrix}-CO-\\-CO-\end{matrix}C_6H_3-SO_3Na + 3NaOH \rightarrow$$

$$C_6H_4\begin{matrix}-CO-\\-CO-\end{matrix}C_6H_2\begin{matrix}ONa\\-ONa\end{matrix} + Na_2SO_3 + H_2O + 2H$$

Nascent hydrogen is obtained during the reaction; this reduces the alizarin to a leuco compound and also causes a poor yield of the dyestuff. The oxidising agent is added to prevent nascent hydrogen from being formed, thus obtaining a good yield of the dyestuff.

$$C_6H_4\begin{matrix}-CO-\\-CO-\end{matrix}C_6H_2\begin{matrix}OH\\OH\end{matrix}$$

Alizarin P (B.A.), **VI** (B.), **IE** (By.), **No. I** (M.), etc., is obtained when anthraquinone-β-sulphonic acid is heated with alkali, etc. It is however usually obtained along with other oxy derivatives of anthraquinone. During the sulphonation of anthraquinone small quantities of disulphonic acid are also obtained, these giving trioxy anthraquinones by alkali fusion.

By further sulphonation of anthraquinone a mixture of 2.6- and 2.7-anthraquinone sulphonic acids is obtained. These compounds may be separated by crystallisation of their sodium salts. They give, on fusing with alkali, **Flavopurpurin** or **Alizarin YCA** (B.A.), **GI, RG** (B.), **VG, XG** (By.), etc.

$$HO\,C_6H_3\begin{matrix}-CO-\\-CO-\end{matrix}C_6H_2\begin{matrix}OH\\OH\end{matrix}$$

and **Anthra-** or **Iso-purpurin** or **Alizarin SC** (B.A.), **SX, GD** (B.), **RF, WR** (By.), etc.

```
                    OH
HO/\-CO-/\OH
  |  |-CO-|  |
   \/       \/
```

Purpurin (B.A.) (B.), or 1 . 2 . 4-trioxyanthraquinone is obtained by oxidation of alizarin in sulphuric acid solution by manganese dioxide.

Alizarin Bordeaux B (By.), etc., is obtained by oxidation of alizarin with strong fuming sulphuric acid, giving a sulphuric ester which is hydrolysed by strong sulphuric acid (80 °/o). The aluminium lake is bordeaux and the chrome lake violet blue.

```
OH        OH
/\-CO-/\OH
|  |-CO-|  |
\/       \/
OH
```

Alizarin Cyanin R, 2R, NS, WRR (By.), etc., is obtained by oxidation of alizarin bordeaux with manganese dioxide similarly to the oxidation of alizarin to purpurin. This dyestuff gives a violet aluminium lake and a blue chrome lake.

```
OH        OH
/\-CO-/\OH
|  |-CO-|  |
\/       \/
OH        OH
```

Anthracene Blue WR (B.).

```
   OH        OH
   /\-CO-/\OH
   |  |-CO-|  |
HO \/       \/
   OH        OH
```

1 . 5-dinitroanthraquinone similarly to 1 . 5-dinitronaphthalene gives on treatment with a solution of

sulphur sesquioxide an oxy-imido compound which, on hydrolysis with ordinary sulphuric acid, gives the hexahydroxyanthraquinone shown above.

By increasing the number of oxy groups in anthraquinone, the polygenetic nature of the resulting dyes becomes gradually diminished.

Nearly all of the above dyestuffs of the anthraquinone series are converted into acid mordant dyes by sulphonation with fuming sulphuric acid.

Alizarin Red S (By.), **WS** (M.), etc., **Alizarin Powder SA**, **Alizarin Carmine** (B.A.) is the product obtained when alizarin is sulphonated with fuming sulphuric acid at 170° C.

OH
—CO— OH
—CO— SO_3Na

Erweco Alizarin Acid Red BS (W.) is a mixture of the sodium salts of alizarin 5- and 8-monosulphonic acids. It is obtained by sulphonation of alizarin with fuming sulphuric acid in presence of mercury. Alizarin 3.5- and alizarin 3.8-disulphonic acids are obtained, which, on hydrolysis, give monosulphonic acids.

Anthracene Brown (R.H.), (B.A.), **W**, **WR** (B.G., etc.), or **Alizarin Brown R, W, H** (M.), etc., or 1.2.3-trioxyanthraquinone is obtained by condensation of benzoic and gallic acids in sulphuric acid solution. It is always associated with some 1.2.3.5.6.7-hexa-oxyanthraquinone or rufigallol, which is also known as **Anthracene Brown SW** (B.) or **Alizarin Brown R, S** (M.), etc.

Alizarine Orange AO, AOP (B.A.), **A, W, SW** (B.), **N** (M.), etc., is obtained by the action of nitric acid upon alizarin in suspension in nitrobenzol, acetic acid, ligroin, or sulphuric acid, to which boric acid is added.

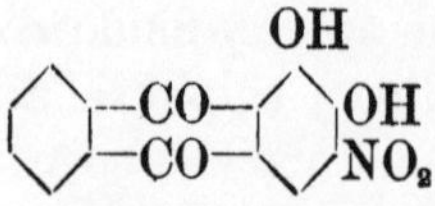

OH
—CO— OH
—CO— NO_2

If the alizarin is benzoylated before nitration, the nitro group enters the alpha position, and on hydrolysis and reduction with sodium sulphide gives **Alizarin Garnet** (M.).

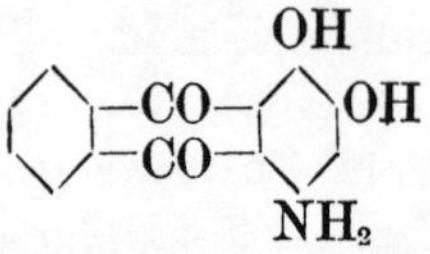

OH
—CO— OH
—CO—
NH_2

Alizarin Blue S, SW (B.), (By.), etc., is the sodium bisulphite compound of **Alizarin Blue ABI** (B.A.), **X, R, WX** (B.), **F** (M.), **GG, XA, WA** (By.), etc.

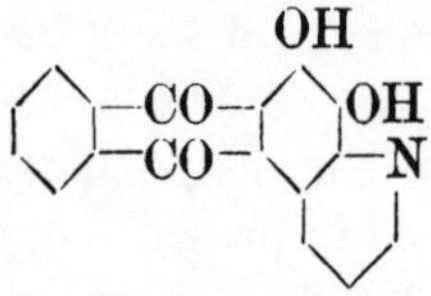

OH
—CO— OH
—CO— —N

It is obtained by the action of glycerin, nitroalizarin, and amidoalizarin in sulphuric acid solution, viz., by the application of Skraup's reaction to nitro derivatives of the anthraquinone dyes.

Alizarin Green S (M.) is a similar substance obtained from α-amidoalizarin, **Alizarin Green X** (B.) and **Alizarin Indigo Blue** (B.) are corresponding compounds obtained from 3-nitro-**Alizarin Bordeaux**, and treatment of this compound with sulphuric acid at 200°C. finally converting into bisulphite compounds, while **Alizarin Black P** (M.) is obtained from 3-nitro-flavopurpurin.

On account of the insolubility of these dyes they are marketed as 20 °/₀ pastes. Soluble powders are obtained by formation of an addition product with two molecules of sodium bisulphite.

Most of the following derivatives of aminoanthraquinones are after-chrome mordant dyes; they may, however, be applied also as simple acid colours.

Alizarin Irisol D, R (By.), etc., is obtained by heating p-toluidine with quinizarin (1.4-dioxy anthraquinone) followed by sulphonation.

```
                OH
   /\ --CO--/\
  |  |      |  |
   \/ --CO--\/
                NH
                |
               /\ - SO3Na
              |  |
               \/
               CH3
```

Anthraquinone Violet (B.) is obtained from 1.5-diamidoanthraquinone by heating with p-toluidine followed by sulphonation of the product thus obtained.

Alizarin Sky Blue B (By.) is obtained by condensation of p-toluidine with dibrom-α-amidoanthraquinone followed by sulphonation.

```
              NH2
   /\ --CO--/\ --Br
  |  |      |  |
   \/ --CO--\/
              |
              NH . C6H3 . CH3 . SO3Na
```

Alizarin Saphirol B (By.) is obtained by sulphonation of 1.5-dioxyanthraquinone (Anthrarufin) followed by nitration and reduction.

```
         NH₂        OH
        /\        /\
       |  |-CO-|  |SO₃Na
NaO₃S  \/-CO- \/
         OH        NH₂
```

Various brands of this dye are used in dyeing mode shades on wool, which are very fast so long as chlorides (NaCl) are absent.

Alizarin Cyanin Green E, G (By.), etc., is obtained by sulphonation of the product produced by heating quinizarin with p-toluidine.

Acid Alizarin Green G (M.) results when dinitro-anthrachrysone disulphonic acid is reduced in alkaline solution with sodium sulphide.

```
          SH        OH
     HO /\        /\ -SO₃Na
  NaO₃S |  |-CO-|  |
        \/ -CO- \/ -OH
          OH        SH
```

See also "The Present Condition of the Chemistry of Anthraquinone," by R. E. Schmidt, *Jour. Soc. Chem. Ind.* 1914, p. 1039.

CHAPTER XXIV

SULPHIDE DYESTUFFS

THE first member of this series, **Cachou de Laval**, an impure and unstable brown dyestuff, was prepared in 1873 by Croissant and Bretonnière by heating vegetable substances, such as sawdust, starch, straw, along with sodium hydroxide and sulphur or sodium sulphide and sulphur. Later it was found that animal and human excrement also gave similar results by fusion with sodium sulphide and sulphur. Cachou de Laval, it was found, acted as a mordant for fixing the basic dyestuffs and also giving different shades (saddenings) by after-treatment with iron or copper mordanting salts. The property of fixing basic dyestuffs is a property possessed by nearly all sulphide dyestuffs, many also having the power of fixing acid and direct cotton dyestuffs, properties of great value in cotton dyeing.

In 1893 R. Vidal produced a sulphide black, **Vidal Black** (P.), from para-amido-phenol, to which he gave the following formula:

```
          H                       H
          N                       N
                    -S-
HO-          -S-           -S-          -OH
                     N
                     H
```

and at about the same period R. Bohn discovered **Fast Black B** (B.), which was obtained from 1.8-dinitronaphthalene. This was the opening of the sulphide era, and during the following ten years almost every organic substance was subjected to sulphide condensation. The varying proportions of sulphur to sodium sulphide, the proportion of the sodium polysulphide produced, the reaction temperature, and the after-treatment of the reaction mass, all play important parts.

The constitution of the sulphide dyestuffs is not yet known. They appear to be thiazine derivatives or similarly constituted polysulphides, possessing the property of dissolving in sodium sulphide solution, the dyestuff being converted by alkaline reduction into the leuco form, and in this form it is absorbed by vegetable fibres and re-oxidised on the fibre, this being aided in some cases by after-treatment with chrome or copper salts.

The sulphide, unlike the vat dyes, re-oxidise to some extent during dyeing.

The sulphide dyes are very insoluble amorphous bodies, giving on reduction thio-oxy derivatives, which are soluble in alkali. Green and Meyenberg found that by oxidising a mixture of a para-diamine or a para-amido-phenol and a large excess of sodium thiosulphate with cold potassium bichromate, a di- or tetra-thio-sulphonic acid is obtained.

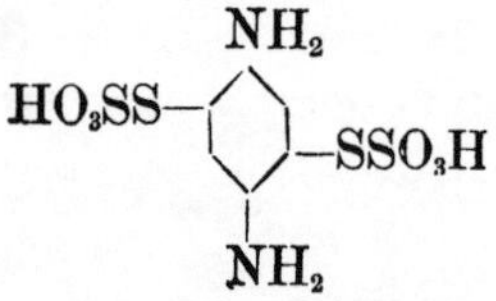

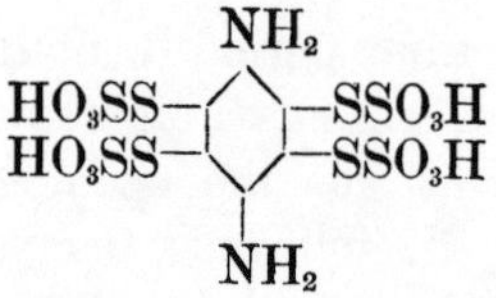

$$NH_2$$
$$HO_3SS-C_6H_2-SSO_3H$$
$$NH_2$$

$$NH_2$$
$$HO_3SS- \quad -SSO_3H$$
$$HO_3SS- \quad -SSO_3H$$
$$NH_2$$

On oxidising either of these compounds with bichromate along with amines, amido-phenols, etc., indamine sulphonic acids are obtained which, on boiling with dilute mineral acids, give products of this series. This method of preparation points to the presence of the thiazine ring in these dyestuffs.

There are practically two methods employed for the preparation of these dyes:

I. Baking.

II. Boiling under reflux condenser.

The latter favours the production of purer products

and minimises the quantity of free sulphur in the final product. The dyes are separated by acid precipitation or air oxidation, the latter method yielding the better product.

Sulphide Yellows.—These dyestuffs are generally obtained from meta-toluylene-diamine, or its derivatives, by heating with sodium polysulphide. The lower the temperature at which the condensation takes place, as a rule, the brighter is the colour of the dyestuff that is produced. In the case of the yellow sulphide dyestuffs the reaction products are often too insoluble to be used directly in the dyebath. By heating with sodium sulphide at about 120° to 125° C. the compounds are rendered more soluble, and may then be evaporated to a solid product or they may be precipitated by acid.

Dehydrothiotoluidine also gives yellows by a sulphide melt treatment, the shade being improved by addition of benzidine or its homologues to the melt.

By raising the temperature for the condensation of m-toluylene-diamine with sulphur, the colour of the final product tends to become more orange or brown.

Meta-toluylene-diamine is employed for the production of **Immedial Yellow D**, **Immedial Orange C** (C.) and **Thion Yellow G** (K.). Thiourea derivatives of the above substance are employed for the production of **Kryogene Yellow** (B.) with or without the addition of benzidine. Para-phenylene-diamine, along with para-amido acetanilide and benzidine, also gives yellow to bronze sulphide dyestuffs, e.g., **Thiophor Yellow Bronze G** (J.).

Sulphide Browns.—Of these dyestuffs only a few are of real commercial value, in spite of the immense number of organic compounds from which they may be obtained.

Kryogene Brown (B.) is obtained from 1 . 8-dinitronaphthalene by treatment with sodium bisulphite, followed by condensation by means of sulphur and sodium sulphide.

Thional Brown (S.) results when certain arylamido derivatives of β-naphthoquinone are condensed at 240° to 280° C. by treatment with sodium polysulphides.

Thiocatechine (P.) is obtained by condensing one part of acetyl-para-phenylene-diamine with two parts of sulphur at 200° to 250° C. As soon as the violent reaction has ended and sulphuretted hydrogen is no longer given off, the mass is allowed to cool. It is soluble in sodium sulphide and dyes cotton catechu brown shades.

Sulphide Reds.—Red is the only colour that is not satisfactorily represented in the sulphide dyestuffs. The red dyestuffs of the azine series, when heated with polysulphide, give those sulphide dyes which most nearly approach to red. Safranine is the one most commonly used. The introduction of copper, nickel and cobalt salts to the melts also tends to produce redder sulphide colours.

Immedial Bordeaux (C.) is obtained from a simple azine, amido-oxyphenazine, by a sulphide melt process.

Sulphide Greens.—Many of the melts which give black sulphide dyestuffs give green dyestuffs when copper salts are added also. Two of the most important intermediate products for the production of these colours are 1-phenylamido-4-p-oxyphenylamidonaphthalene-8-sulphonic acid

HO⟨ ⟩—NH—⟨ ⟩—NH—C_6H_5
⟨ ⟩—SO_3H

and 4-p-oxyphenylamido-1-amidonaphthalene sulphonic acid

HO⟨ ⟩—NH—⟨ ⟩—NH—SO_3H
⟨ ⟩

Sulphide Blues.—The blues are mainly obtained from indophenols. These compounds are easily produced by condensation of para-diamines, or para-amidophenols, with amines or phenols. Both benzene and naphthalene derivatives are largely used for this class. Para-oxy-para-amidodiphenylamine obtained from aniline and para-amidophenol by oxidation with bichromate, or by oxidation of a mixture of phenol and para-phenylene-diamine, and also para-phenylamino-para-oxy-diphenyl-amine obtained from diphenylamine and para-nitroso-phenol, are both extensively used for the manufacture of sulphide blues.

Immedial Sky Blue (C.) is obtained by condensing dimethyl-p-amido-p-oxydiphenylamine with sodium poly-sulphide at 110° to 115°C. by heating under reflux condenser. Its probable formula is

$(CH_3)_2N$⟨ ⟩—N=⟨ ⟩—S—S—⟨ ⟩=N—⟨ ⟩—$N(CH_3)_2$
—S— =O O= —S—

Sulphide Blacks.—A large variety of nitro-compounds may be used for sulphide black melts. 2.4-dinitrophenol, 1.5- and 1.8- or the mixture of these two dinitronaphthalenes, 1.5-dinitroanthraquinone, di-nitrodiphenylamine, together with their oxy and chloro derivatives or corresponding derivatives of diphenylamine, are the main intermediate compounds used for sulphide blacks.

The preparation of sulphide blacks may be best understood, however, from actual examples.

Sulphur Black T (Ber.). 85 parts of crystallised sodium sulphide are dissolved in 100 parts of water; 30 parts of sulphur are dissolved in the above solution by heating on the water bath. To the above solution 20 parts of 2.4-dinitrophenol are gradually added and the reaction mass heated for 20 hours under reflux condenser. The solution is then tested by spotting a drop on filter paper; if the reaction is complete no yellow colour will be seen at the edge of the spot. If a yellow edge is shown boiling is continued until it is no longer produced. The dyestuff is obtained from this solution by precipitation by means of acid or aeration.

Immedial Black V (C.). 20 parts of 2.4-dinitro-4′-oxy-diphenylamine are added slowly to a solution of 16 parts of sulphur in 44 parts of crystallised sodium sulphide. The reaction mass is now heated, in an oil bath, to 140°C. during five hours under reflux condenser. The solution becomes deep blue-black in colour, and about this stage a vigorous evolution of sulphuretted hydrogen takes place. As soon as this has subsided, the solution is diluted and air drawn or blown through it until no more dye is precipitated; it is then filtered, washed, and dried.

Unless carefully manufactured, sulphide dyes, especially blacks and browns, are liable to contain sulphur in a loosely combined form, which oxidises readily to free sulphuric acid, e.g., in the drying stove, or after dyeing when goods are subjected to heat and moisture, thus tendering the fabric. Alkaline impregnation of cloth has done much to diminish tendering from this cause, but still more is due to improvements in the methods of dye manufacture, e.g., the use of air precipitation instead of acid, the accurate adjustment of melt quantities and the replacement of baking methods by condensation in solution. The sulphide dyestuffs find their chief use on vegetable fibres, as the hot alkaline bath, from which they are dyed, acts unfavourably on animal fibres and tissues. A noteworthy fact in connection with the sulphide dyes, which in most respects possesses considerable fastness, is their small resistance to hypochlorites.

The following are among the best known brands of sulphide dyestuffs :—**Auronal** (W-t-M.), **Cross Dye** (R.H.), **Eclipse** (G.), **Immedial** (C.), **Katigen** (By.), **Kryogene** (B.), **Pyrogene** (S.C.I.), **Sulphur** (Ber.), **Thiogene** (M.), **Thion** (K.), **Thional** (S.), **Thionol** (Lev.), **Thiophor** (J.), **Thioxine** (G.E.), **Vidal** (P.).

CHAPTER XXV

VAT DYESTUFFS

UNDER this group are included all those dyes which are insoluble in water, but can be applied to the fibre through the intermediate production of a reduction or leuco compound which is soluble in dilute caustic alkali. The leuco compound is subsequently re-oxidised on the fibre to the original dyestuff. For this purpose a dyestuff must contain the ketonic or thioketonic group, which on reduction gives respectively a phenolic or thiophenolic group capable of solution in alkali. The alkali compound of the phenol or thiophenol is taken up by the fibre and is then re-oxidised by air to the original ketone or thioketone. In many cases it is advisable also to treat the dyed material with some oxidising agent, such as bichrome or perborate, etc.

Indigo, which is the oldest and best-known member of this series, is very fast to light and to most other agents. It is faster however when dyed on wool than when dyed on cotton, wool probably having a more pronounced chemical affinity for indigo than cotton. The fastness to light of indigo is surpassed by many other vat dyes, but such fastness is by no means invariable in the class of vat dyestuffs.

The vat dyestuffs may be subdivided into three classes:

I. The Indigoid Vat Dyestuffs.
II. The Anthraquinone Vat Dyestuffs.
III. The Sulphurised Vat Dyestuffs.

No really sharp class distinction, however, can be drawn in the case of certain dyes of these series.

The Indigoid Class may be applied to wool as their leuco compounds are soluble in very dilute alkali. The amount of alkali permissible without risk of injury to the wool must be exceeded in many cases however, and various "restrainers" have been added in dyeing, e.g., glue, glucose, formaldehyde, and soluble oils.

The Anthraquinone Class are mainly in use for cotton dyeing, and are not generally applied to wool because they usually require too much caustic alkali to dissolve their leuco compounds. They may, however, in certain cases be applied to wool, although great difficulties have to be overcome.

The Sulphurised Class are also more important for cotton dyeing.

The Indigoid Class.—The class of indigo derivatives is characterised by the presence of one or two five-membered rings, in which a ketonic group is present along with either carbon, nitrogen, or sulphur, one of the carbon atoms of this ring being combined by a double bond to a similar carbon or to a ring component carbon atom, as for example to anthracene or naphthalene, provided there is a ketonic group ortho to the double-bonded carbon. Several types are possible therefore as represented by the following chromogenes:

Indigo and derivatives

Thioindigo and derivatives

Ciba Grey etc.

Indirubin etc.

Thioindigo Scarlet etc.

Ciba Scarlet etc.

Helindone Blue or Alizarin Indigo etc.

When synthetic indigo was first placed on the market many virtues were claimed by dyers for natural indigo as compared with the chemically pure synthetic product. The valuable red bloom on indigo navies is more easily obtained with the older dyes, and this was attributed to the natural indigo containing indirubin (indigo red). The natural indigo also contains small amounts of indigo brown, indiglutin, etc. It has been shown, however, that indirubin, on reduction in the indigo vat, gives indoxyl which oxidises later in presence of air, giving indigotin or indigo blue. The indiglutin, a degradation product of indigo, is perhaps the most important impurity apart from clay and gums which may exert a colloidal influence. Owing to the varying quantities of substances other than indigotin present in natural indigo, the shade produced from it varies with different samples, and also varies during dyeing from the same vat. The shade produced by synthetic indigo is always the same, if the same percentage of dye is fixed on the fibre, and the same method of dyeing is employed. The shade of indigo varies, however, with the manner of application and the after-treatment. For example, steaming the dyed goods for a short time makes the blue a little more violet, and also increases its fastness against light.

Indigo (indigotin) is present in both plant and animal life in the form of its glucoside **indican.** This glucoside is formed in small quantity by a process of de-assimilation in the stalk and leaves of *indigofera tinctoria* and *indigofera sumatrana,* which are cultivated chiefly in India and Java. The leaf of the indigo plant contains, on an average, about 0·5 per cent. of colouring matter. Two crops are gathered from the same plants each year, the second crop some two or three months after the first. The cut plant is put into vats and extracted with water, the temperature of which is about 30° to 35° C.; the length of time required for extraction varies from nine to fourteen hours. Towards the end of the extraction marsh gas and hydrogen are evolved, and about this period the fermentation diminishes and the liquid subsides. The liquid is then run off into the beating vats, in which oxygen from the air acts as the oxidising agent and precipitates the indigo from the vat. It is then filtered, boiled with water, pressed, dried in the shade, and cut into blocks.

The amount of indigotin present in natural indigo varies very widely. Java indigo usually contains most, as a rule 70 per cent. and upwards, while good Bengal indigo contains 60 to 65 per cent. and Madras indigo 30 to 40 per cent. As much as 10 per cent. of indirubin is found in some commercial samples.

The shortage of artificial indigo, due to the war, has done much to revive indigo cultivation again, which dwindled rapidly after the synthetic product appeared on the market.

The Badische Anilin und Sodafabrik manufactured indigo in 1897 by the Heumann method which consists in the fusion of phenylglycine with caustic soda, followed by oxidation of the alkali melt, after solution in water, with air.

Phenylglycine by loss of water gives indoxyl thus:

$$C_6H_4\begin{matrix}\diagup NH \diagdown \\ \diagdown HOCO \diagup\end{matrix}CH_2 \rightarrow C_6H_4\begin{matrix}\diagup NH \diagdown \\ \diagdown COH \diagup\end{matrix}CH + H_2O$$

indoxyl on oxidation, e.g., with air, gives indigo:

$$2 \cdot C_6H_4\begin{matrix}\diagup NH \diagdown \\ \diagdown COH \diagup\end{matrix}CH + O_2 \rightarrow$$

$$C_6H_4\begin{matrix}NH \\ CO\end{matrix}\!\!>C=C<\!\!\begin{matrix}NH \\ CO\end{matrix}C_6H_4 + 2H_2O$$

The yield by this method was extremely poor, but by replacing the phenylglycine by its ortho carboxylic acid a fairly good yield is obtained. The phenylglycine-o-carboxylic acid is obtained either from naphthalene or from o-chlorobenzoic acid. Naphthalene is oxidised by sulphuric acid in presence of mercury to furnish phthalic anhydride, and the latter is converted by ammonia into phthalimide $C_6H_4\langle{}^{CO}_{CO}\rangle NH$, which by treatment with sodium hypochlorite gives anthranilic acid $C_6H_4\langle{}^{NH_2}_{COOH}$, this being converted into phenylglycine-o-carboxylic acid by condensation with chloroacetic acid. The o-chlorobenzoic acid produces the same compound by heating with glycine in presence of copper salts.

A better yield of indigo is obtained by condensation of phenylglycine by heating with sodamide and caustic soda, or by heating methylanthranilic acid with sodamide and caustic soda. The use of sodamide for indigo manufacture was discovered by the Deutsche Gold u. Silberscheide Anstalt and purchased by Meister, Lucius and Brüning. The sodamide may be produced in the reaction mixture of sodium and the sodium salt of phenylglycine by action of ammonia.

Sodium anilide (C_6H_5NHNa) also finds application in the indigo synthesis, this compound increasing the yield of indigo similarly to sodamide (S. C. I.).

Whatever method is employed it is essential to carry out the condensation in absence of air to prevent the formation of isatin $C_6H_4\langle{}^{CO}_{NH}\rangle CO$, an oxidation product of indigo, which causes the formation of indirubin.

Indigotin is also placed on the market in a reduced form, **Indigo White**, which may be prepared from indigo by reduction with iron powder and caustic soda, or by glucose and caustic soda, such as **Indigo White BASF** (B.) or **Indigo MLB/W** (M.).

When indigo is reduced in the vat two atoms of hydrogen are taken up by the molecule of dyestuff, the following compound

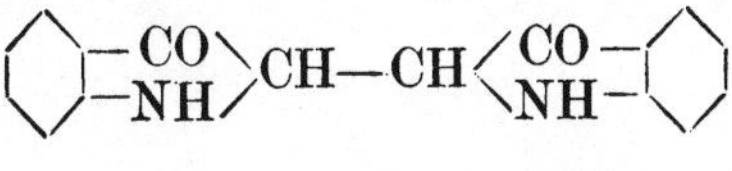

$$\bigcirc\begin{matrix}-CO\\-NH\end{matrix}\!>CH-CH<\!\begin{matrix}CO-\\NH-\end{matrix}\bigcirc$$

probably being produced. This transforms into **Indigo White**

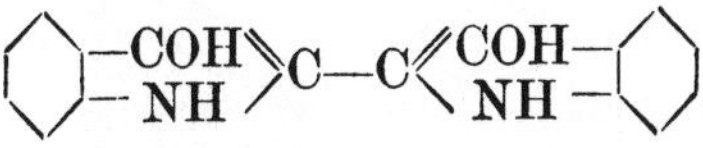

$$\bigcirc\begin{matrix}-COH\\-NH\end{matrix}\!>C-C<\!\begin{matrix}COH-\\NH-\end{matrix}\bigcirc$$

the leuco compound dissolving in the alkali present to give a yellowish green vat.

In order that synthetic indigo may be readily reduced it is placed on the market in as fine a state as possible, and in many cases compounds are added to preserve the indigo in colloidal form during oxidation, e.g., benzylaniline sulphonic acid (B., 1912).

Indigo Salt T (K.) is obtained by the action of dilute caustic soda upon a solution of ortho-nitrobenzaldehyde in acetone.

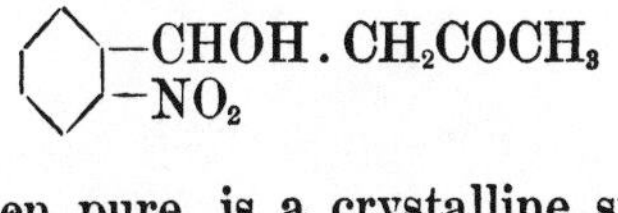

$$\bigcirc\begin{matrix}-CHOH\,.\,CH_2COCH_3\\-NO_2\end{matrix}$$

Indigo, when pure, is a crystalline substance with a coppery hue subliming to give violet vapours on heating. This property of sublimation is common to indigoid vat dyestuffs, and is useful as a means of distinguishing from anthraquinone and sulphurised vat dyestuffs which—with the exception of **Anthraflavone** (B.)—do not sublime. Another means of distinguishing lies in the different coloured leuco compounds obtained in the vat on reduction. While insoluble in water and common solvents, the indigoids are, generally speaking, more easily dissolved than anthraquinone and sulphurised vat dyestuffs. Thus indigo itself is soluble in boiling glacial acetic acid, pyridine, phenol, benzaldehyde, nitrobenzene, etc., but only sparingly soluble in any of these in the cold.

On dissolving indigo in concentrated sulphuric acid at 70°C. it becomes converted into a mono-sulphonic acid (—SO_3H para to NH—). This, as sodium salt, constitutes the commercial red or purple **Indigo Extract**. The disulphonic acid obtained by longer sulphonation is more used—**Indigo Carmine.** It finds some use as an acid dye of no particular fastness.

Indigo is readily converted into isatin by treatment with oxidising agents, e.g., nitric acid, chromic acid, etc. This is made use of technically in discharging indigo dyed cloth, and in the estimation of indigo (as sulphonate) by titration with permanganate.

[The common facts relating to the chemistry of indigo are to be found in all organic chemistry text-books, and for that reason a somewhat curtailed treatment has been accorded here.]

Derivatives of Indigo.—The halogen derivatives are perhaps the most important. These may be produced from halogen compounds or the indigo may be halogenated, the halogenation being carried out in a solution or in presence of nitrobenzene, dichlorobenzene, concentrated sulphuric acid or chlorosulphonic acid. By progressive bromination of indigo it is possible to introduce from one to six atoms of bromine into the indigo molecule.

The indigo derivatives are named according to the positions of the groups, the positions being numbered as follows:

```
    4   3                    3′    4′
   / \ /CO\  2   2′  /CO \ / \
 5|   |     \C=C/       |   |5′
 6|   |     /    \      |   |6′
   \ / \NH/        \NH/ \ /
    7   1            1′    7′
```

Bromine first enters into position 5 forming a mono- or di-bromo-derivative continuing through the following series: 55′7 to 55′77′, finally to 5′57′74′ and 5′57′74′4. The 66′ compound cannot be obtained by bromination of indigo, but is produced from 4-bromo-2-amino-benzoic acid. The 66′ dibromo-indigo was found by Friedlaender to be one of the dyestuffs present in the shell-fish *murex brandaris*, from which **Tyrian Purple** was obtained by the ancients.

The chloro indigos are obtained from the halogen derivatives of phenylglycine.

The halogenated indigos are brighter in shade and faster to bleaching agents than indigo itself, but by increasing the substitution the relative tinctorial power is decreased.

Homologues of indigo do not appear to have any special value; a 77′ dimethyl indigo, **Indigo MLB/T** (M.) or **Indigo pure BASF/G** (B.), is, however, manufactured. It dyes greener shades, which are faster to chlorine than indigo.

Amino derivatives of indigo give brown vat dyes;

a brominated diamino-indigo, **Ciba Brown R** (S.C.I.), is on the market. It dyes cotton and wool reddish brown, which is fast to light and washing but not to bleaching.

Naphthalene indigos have been obtained from α- and β-naphthylamine; they are green dyestuffs, but are comparatively fugitive and therefore are of little value. The beta-compound, however, gives valuable vat dyes on bromination. **Ciba Green G** (S.C.I.) and **Helindone Green G** (M.) belong to this class.

```
Br                        B
 -NH  NH-
  |    |
CO-C=C-CO
```

The naphthalene indigos show less disposition to sublime on heating than the simple indigos. **Ciba Green G** when heated gives some reddish violet vapours, but most of the dyestuffs carbonise without subliming.

Indigo Yellow 3G (S.C.I.) is prepared by heating indigo with benzoyl chloride in nitrobenzene solution in presence of copper powder. It gives a bluish-red vat on alkaline reduction. It is of particular interest, since it may be dyed along with indigo from the same vat giving uniform green shades. Generally a mixture of indigoid dyes gives uneven shades due to unequal affinity of these dyes for the fibre and the special amounts of alkali and particular temperatures required. It has most probably the following constitution:

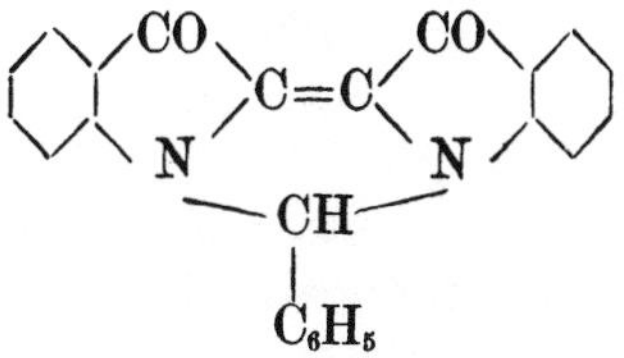

```
  /\  /CO\      /CO\  /\
 |  ||     >C=C<    ||  |
  \/ \ N  /      \  N /\/
        \—— CH ——/
            |
           C6H5
```

Most natural indigos contain **Indirubin.** This compound is obtained by condensation of isatin with indoxyl,

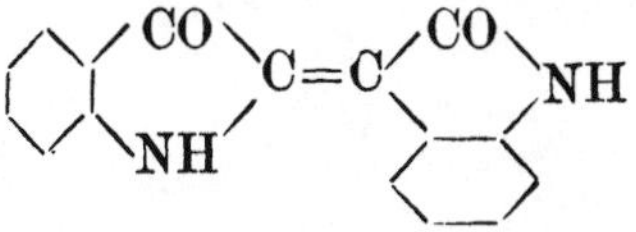

```
  /\  / CO\      /CO\
 |  ||     C=C<      NH
  \/ \NH  /      \   /
                 /——\
                 \__/
```

and is also obtained during the production of synthetic indigo when air is admitted to the caustic melt. Indirubin, when vatted, is to a large extent converted into indigo-blue, this is accompanied with loss of dyestuff. The amount of indirubin dyed and fixed on wool is therefore small and also it is of little value, as it gives shades loose to washing. Certain derivatives of indirubin are sufficiently fast, however, to be employed commercially as vat dyestuffs, notably halogen derivatives. By bromination the fastness to washing is improved. **Ciba Heliotrope B** (S.C.I.) is a tetrabrominated indirubin giving a yellowish olive vat.

In 1905 Friedlaender discovered **Thioindigo Red B** (K.), this being the first of a large series of vat dyes. It is similar in constitution to indigo, the NH group being replaced by sulphur. Its derivatives are named similarly to those of indigo.

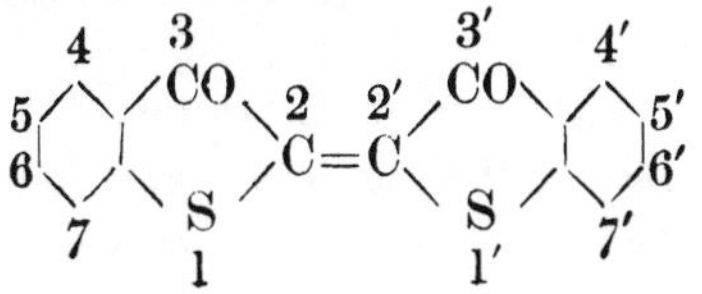

```
    4   3              3'    4'
 5 /\  /CO\ 2      2' /CO\  /\ 5'
  |  ||     >C=C<         ||  |
 6 \/ \  S /      \  S   / \/ 6'
    7    1           1'     7'
```

It is termed bisthionaphtheneindigo and is obtained from thiosalicylic acid, which, on heating with sodium chloroacetate, gives phenylthioglycol-ortho-carboxylic acid. This, on heating, gives thioindoxyl, which may be oxidised to **Thioindigo Red** (K.) (cf. indigo synthesis).

$$C_6H_4\begin{matrix}-SH\\-COOH\end{matrix} \rightarrow C_6H_4\begin{matrix}-S-CH_2-COOH\\-COOH\end{matrix}$$

$$\rightarrow C_6H_4\begin{matrix}\diagup S \diagdown\\ \diagdown COH \diagup\!\!\!\!\diagup\end{matrix}C-COOH \rightarrow C_6H_4\begin{matrix}\diagup S \diagdown\\ \diagdown CO \diagup\end{matrix}C=H_2$$

$$\text{or } C_6H_4\begin{matrix}\diagup S \diagdown\\ \diagdown C(OH) \diagup\!\!\!\!\diagup\end{matrix}CH$$

Thioindoxyl on oxidation, preferably with potassium ferricyanide, gives **Thioindigo Red.**

Many derivatives of **Thioindigo Red** have been prepared, introducing a variety of shades.

Helindone Grey BR (M.) is a dichloro-diamido-thioindigo dyeing from a yellow vat.

Helindone Violet (M.) is dichlorodimethyl-dimethoxy-thioindigo.

Helindone Scarlet S (M.) is diethylthio-oxy-thioindigo.

Ciba Bordeaux B (S.C.I.) is 5.5′-dibrom-bisthionaphtheneindigo. It forms a yellowish orange vat on reduction.

By the condensation of the thionaphthene ring with the indoxyl or indol ring a variety of vat dyes are obtained, these being the thionaphthene-indol indigos;

the combination of the two ring systems may be at the 2 or 3 positions both in the indol ring and the thio-naphthene ring. Oxythionaphthene or its derivatives is condensed with isatin or its derivatives, or with alpha-isatin derivatives.

One of the most important is **Ciba Red G** (S.C.I.) or **Thioindigo Scarlet G** (K.), which is obtained by the condensation of isatin with alpha-oxy-thionaphthene followed by bromination.

```
      4   3               2
   5 / \ /CO\  2   3   /CO\
   6|   |     \C=C       NH 1
     \ / \ S /      \  /
      7    1     4 <    > Br 7
                   5 Br  6
```

It may be termed 2-thionaphthene-5.7-dibrom-3-indol indigo. It dyes from a pale yellow vat and is a valuable dyestuff, being fast to light and bleaching.

Other dyestuffs of this series are:

Ciba Violet B (S.C.I.) or tribrom-2-thionaphthene-2-indol indigo.

Ciba Grey G (S.C.I.) or brom-2-thionaphthene-2-indol indigo.

Acenaphthoquinone also condenses with indoxyl or oxythionaphthene or their derivatives to produce vat dyes. One of good fastness to almost all agencies is **Ciba Scarlet G** (S.C.I.) or 2-thionaphthene-acenaphthene indigo.

```
                    CO—
   / \ /CO\          |
  |   |     C=C——
   \ / \ S /
```

Vat dyes may also be obtained by the condensation of isatin or its brom derivatives with naphthol or anthranol or certain derivatives of these compounds.

Alizarin Indigo 3R (By.) or 2-naphthalene-2-indol indigo gives a yellow vat on reduction. It is obtained when dibromisatin is condensed in benzene solution with alpha-naphthol.

Helindone Blue 3GN (M.) is obtained by the condensation of oxyanthranol with isatin anilide.

Alizarin Indigo G (By.) is a brominated 2-anthracene-2-indol indigo.

Anthraquinone Vat Dyes.

The dyes of this series may be roughly divided into three types: a ring system in which nitrogen is absent, a ring system in which nitrogen is present (mainly dyes obtainable from amino anthraquinones), and sulphurisation products of anthraquinone and its derivatives. Many of these dyes are particularly interesting on account of their high molecular weight and their complex structural formula. Many coloured bodies derived from anthraquinone and such of its derivatives as furnish a leuco body in the hydrosulphite vat, do not possess the necessary affinity for the fibre.

The simpler vat dyestuffs of this series may be compared with the parent body anthraquinone; this body is insoluble in alkalis, but may be reduced by alkaline reducing agents giving oxanthranol.

$$C_6H_4{<}\begin{matrix}CO\\CO\end{matrix}{>}C_6H_4 + 2H \rightarrow C_6H_4{<}\begin{matrix}CHOH\\CO\end{matrix}{>}C_6H_4$$

The oxanthranol is soluble in alkalis, forming a blood red solution; this compound on exposure to air gives the original quinone again. Compounds similar to this, if of sufficient tinctorial power, may therefore be used as vat dyes.

Anthraflavone G (B.) is a stilbene derivative obtained by heating β-methylanthraquinone with alcoholic potash from 150° to 170° C. It dyes cotton a fast greenish yellow from a dark reddish-brown vat.

$$C_6H_4{<}\begin{matrix}CO\\CO\end{matrix}{>}C_6H_3-CH=CH-C_6H_3{<}\begin{matrix}CO\\CO\end{matrix}{>}C_6H_4$$

Indanthrene Gold Orange G (B.) is obtained from 2.2′-dimethyl-1.1′-dianthraquinonyl by heating with or without dehydrating agents. It gives a magenta coloured vat.

$$\begin{matrix}C_6H_4{<}\begin{matrix}CO\\CO\end{matrix}{>}C_6H_2-CH_3\\ |\\ H_3C-C_6H_2{<}\begin{matrix}CO\\CO\end{matrix}{>}C_6H_4\end{matrix} \rightarrow \text{(polycyclic diketone, O and O)}$$

In this and certain other formulae for vat dyes the quinone form =O is written instead of the ketonic —CO— in order to simplify complex formula.

Indanthrene Gold Orange R (B.) is a chloro derivative of the above, and **Indanthrene Scarlet G** (B.) a bromo derivative of it; it gives a reddish-violet vat.

Indanthrene Dark Blue BO (B.) is another example of the peculiar complex condensation that may be carried out with anthracene derivatives.

Benzanthrone, obtained by heating anthranol, glycerine and sulphuric acid, is heated with caustic potash, whereby two molecules condense together.

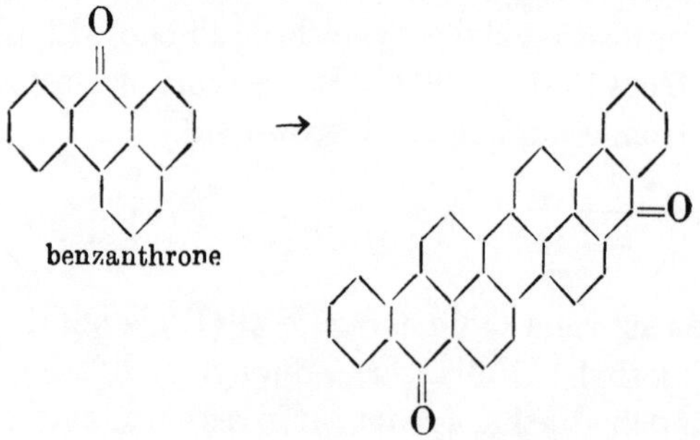

benzanthrone →

Indanthrene Dark Blue BO (B.)

Indanthrene Green B (B.) is a nitration product of the previous dyestuff; it is particularly interesting, since by the action of oxidising agents upon cotton dyed with this colour a fast black is obtained, namely, **Indanthrene Black B** (B.).

Indanthrene Violet R extra (B.) is isomeric with **Indanthrene Dark Blue BO** (B.); it is obtained by a caustic potash melt of halogen derivatives of benzanthrone.

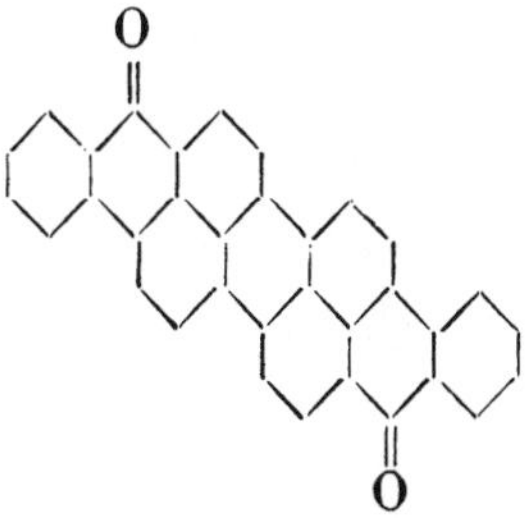

It was found at the Bayer Farben Fabrik that simpler compounds, benzoylated oxy- or amido- anthraquinones, had the properties of vat dyes.

Algol Yellow WG (By.) is benzoyl-1-amidoanthraquinone.

Algol Scarlet G (By.) is benzoyl-1-amido-4-methoxy anthraquinone.

Algol Red 5G (By.) is dibenzoyl-1.4-diamidoanthraquinone.

Algol Red R extra (By.) is dibenzoyl-1.5-diamido-8-oxy anthraquinone, and gives a red vat.

The first two members of the anthraquinone vat series to be discovered contain nitrogen, and were obtained from β-amidoanthraquinone by fusion with caustic alkali. At temperatures from 200° to 250° C. a blue dye is obtained, and at 330°—350° C. a yellow one. These two dyes were discovered in 1901 by R. Bohn at the Badische Anilin und Sodafabrik.

Indanthrene Blue R (B.) is one of the oldest of the anthraquinone vat dyes, and is still the most important and the most largely used of these colours. It is obtained by heating β-amidoanthraquinone with caustic potash and potassium nitrate. The melt is dissolved in water and the dyestuff which separates is filtered off. It is purified by reduction to the leuco form and separation of the sodium salt from caustic soda solution, this being then well washed with alkaline bisulphite solution.

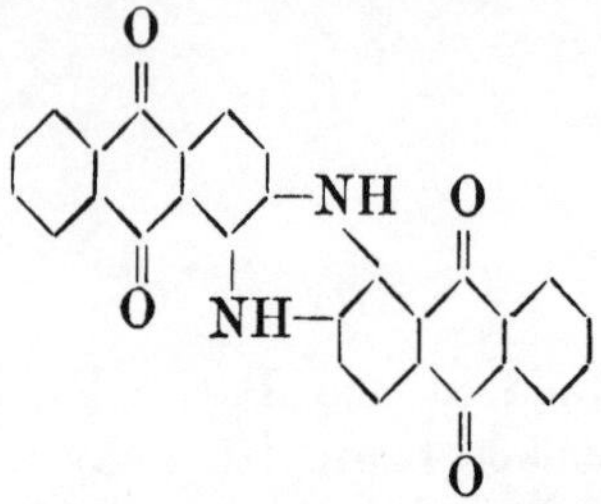

It would be termed as a leuco-azine N-dihydro-1.2-1′.2′-anthraquinone azine. It dyes bright blue shades from a dark blue vat.

Algol Blue K (By.) is the N-dimethyl derivative of the above compound.

Algol Blue 3G (By.) is 4.4′-dioxyindanthrene.

Indanthrene Yellow G (B.) is obtained similarly to Indanthrene Blue R, except that the condensation temperature is much higher. It may also be obtained by heating β-amidoanthraquinone with antimony pentachloride in nitrobenzene solution.

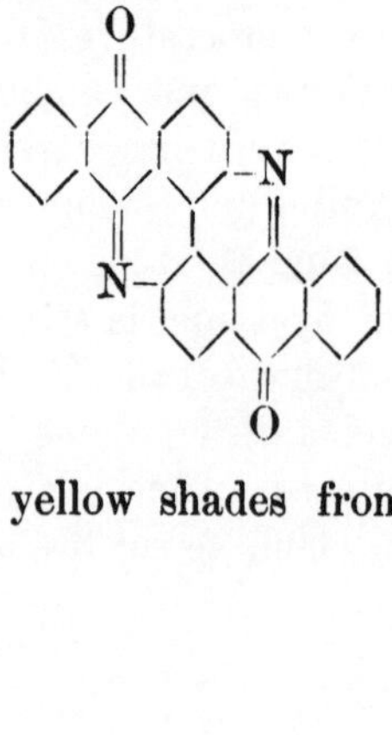

It dyes bright yellow shades from a dark bluish violet vat.

Helindone Yellow 3GN (M.) is 2.2′-dianthraquinonyl urea.

Imides of the anthraquinone series form a large class of vat dyes of which the following are a few examples.

Algol Red B (By.) is β-anthraquinone-α-anthra-N-methylpyridonamine.

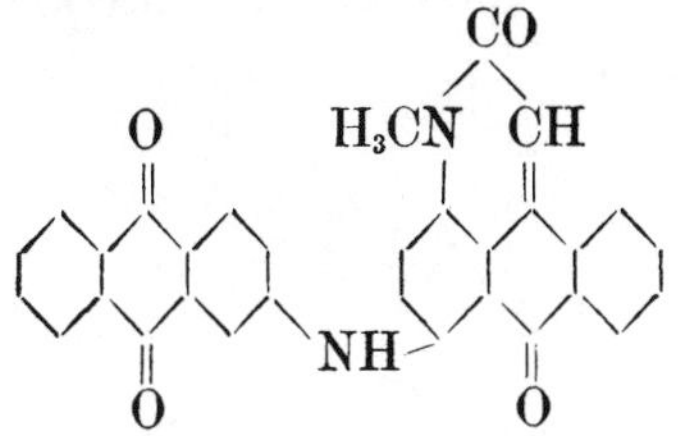

CO
H_3CN CH
O
NH
O O

Indanthrene Bordeaux B (B.) is dichloro-di-α-anthraquinonyl-2.7-diamido-anthraquinone.

O O
Cl —Cl
O NH NHO
—CO—
CO

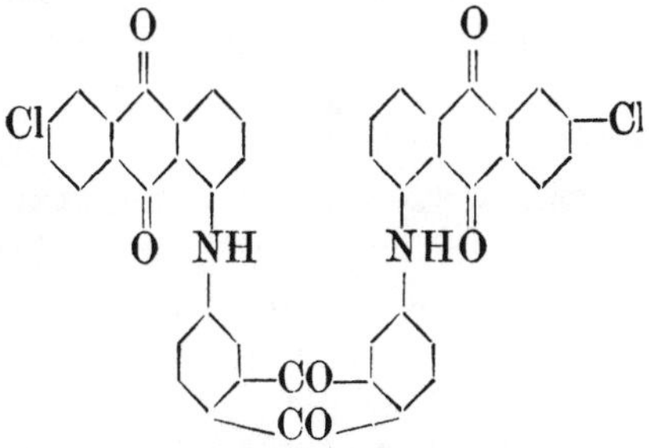

Algol Bordeaux 3B (By.) is 4.4′-dimethoxy-di-α-anthraquinonyl-2.6-diamido-anthraquinone.

Vat dyes of the acridone series.—These dyestuffs are obtained by condensing chloro anthraquinone derivatives with anthranilic acid.

Indanthrene Red BN extra (B.) is an anthraquinonenaphthacridone.

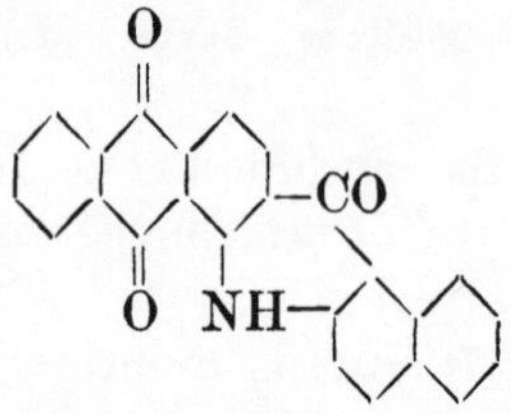

Indanthrene Violet RN extra (B.) is an anthraquinonediacridone.

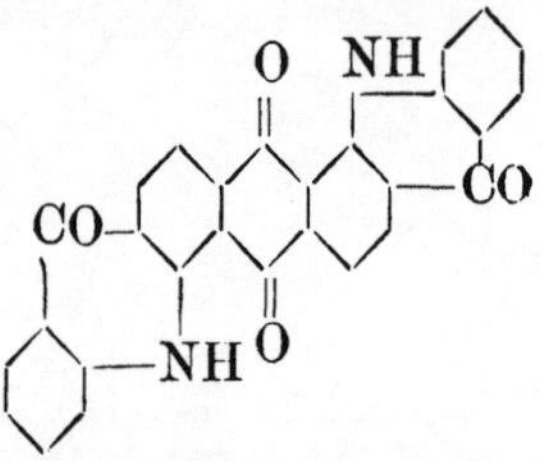

Sulphurised Vat Dyes.—These are obtained from anthracene and many of its derivatives or condensation products by heating with sulphur to a high temperature.

Indanthrene Olive G (B.) is produced by heating one part of anthracene (96 to 98 %) with three parts of sulphur to 250° C. until sulphuretted hydrogen is no longer given off. It gives a dark violet vat which dyes olive shades fast to light and washing.

Cibanone Blue 3G (S.C.I.) is a sulphur melt product of methylbenzanthrone.

Cibanone Black B (S.C.I.), **Cibanone Green B**, **Cibanone Orange R**, and **Cibanone Yellow R** (S.C.I.) are all vat dyes of this series.

By the condensation of indophenols of carbazol, or N-substituted carbazols with sodium polysulphide, vat colours are obtained.

Hydron Blue R (C.) is obtained from the indophenol of carbazol and sodium polysulphide. It is one of the vat dyes most used on cotton.

—NH—
—N= =O

The addition of copper salts to the melt gives **Indocarbon S** (C.).

(For reference:
Barnett, "Chemistry of Vat Dyes," *J. Soc. Dyers*, 1913, p. 183.
Vlies, "Recent Progress in Colouring Matters," *J. Soc. Dyers*, 1913, p. 316; 1914, pp. 22 and 29.)

CHAPTER XXVI

NATURAL DYESTUFFS

MOST natural dyestuffs are mordant dyestuffs and occur as glucosides in plants. Among these, the yellow dyestuffs which have been identified are mainly oxy-derivatives of flavone[1] or flavanol.

Young Fustic contains a tri-oxy-flavanol termed

OH
HO O C— OH
CO—COH

Fisetin. It is obtained from the wood of the sumach tree (*rhus cotinus*), a small tree indigenous to Southern Europe and the West Indian Islands. It is polygenetic and may be applied to the dyeing of wool. On account

[1] Flavone itself occurs almost pure as the mealy deposit found on the leaves of many plants of the primula class.

— O —C—C_6H_5
—CO—CH

of the shades obtained from it lacking permanency it is little used.

Old Fustic is the wood of the tree *chlorophora tinctoria*, previously termed *morus tinctoria*. It is principally used in the dyeing of wool. Fustic consists mainly of two dyestuffs:

Morin HO, O, OH, C–, OH, CO–COH, OH

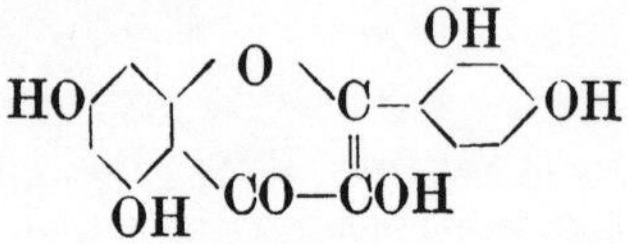

and **Maclurin** OH, CO, HO, OH, OH, OH

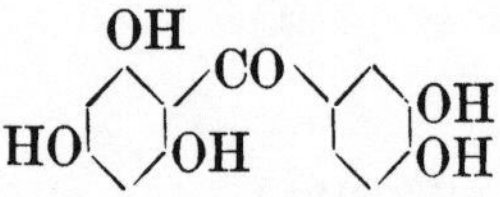

These are polygenetic, giving yellow shades with aluminium, tin and chrome, and olive with iron and copper mordants, although morin does not contain ortho-oxy-groups (see p. 110).

The wood of the Osage Orange tree found in the United States behaves like Fustic in dyeing.

Patent Fustin is a colouring matter placed on the market which is mainly disazobenzene maclurin. It is obtained by extracting Old Fustic with boiling water, the solution filtered from the morin and its calcium salt, which separate on cooling, the filtrate neutralised with sodium carbonate and diazobenzene sulphate added until no further precipitation takes place; this precipitate is collected, washed, and sold in paste form. The disazobenzene maclurin has the following formula:

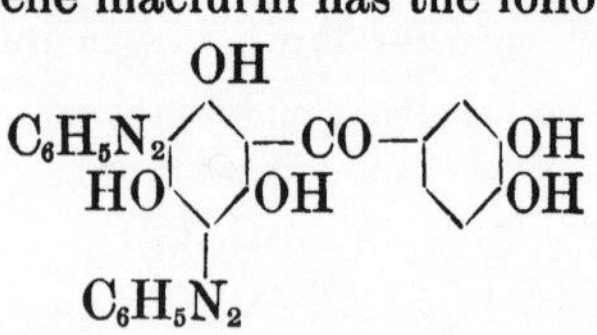

OH, $C_6H_5N_2$, –CO–, OH, HO, OH, OH, $C_6H_5N_2$

Persian Berries consists of the dried unripe fruit of various species of *rhamnus.* It consists mainly of two flavanol derivatives, **rhamnetin** and **rhamnazin**, and small quantities of **quercetin.** It is employed for dyeing paper, leather, wool, in wool and calico printing for yellow shades, and for colouring foodstuffs.

Turmeric is contained in the tuber of *curcuma tinctoria.* It is now mainly employed for tinting and flavouring butter, cheese, confectionery, and for tinting lakes, fats and varnishes.

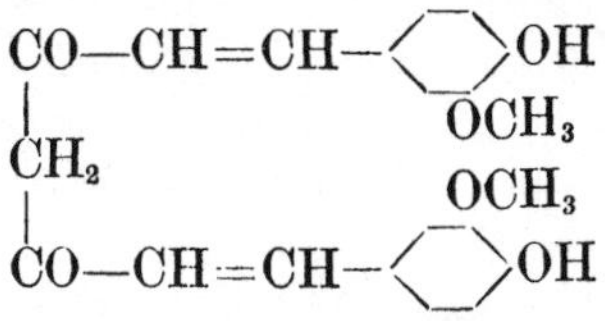

$$
\begin{array}{l}
CO{-}CH{=}CH{-}C_6H_3(OH)(OCH_3) \\
| \\
CH_2 \\
| \\
CO{-}CH{=}CH{-}C_6H_3(OH)(OCH_3)
\end{array}
$$

It is interesting as it is directly absorbed by cotton, and may be dyed in presence of a little acetic acid or alum. It is feeble to light and alkalis, acquiring a brownish-red tint by treatment with the latter.

Annatto is obtained from the fleshy pulp surrounding the seeds of *bixa orellana.* It is mainly cultivated in Central America, in Cayenne, in Brazil, and in India, the latter giving the best seeds. The colouring matter is termed **bixin** $C_{28}H_{34}O_5$. Owing to the fugitive nature of the red and orange shades obtained by it, its employment is very limited.

It is used for colouring butter, cheese, margarine, soups, etc.

Quercitron Bark is the inner bark of a species of oak, *quercus discolor* ; it is indigenous to the Central and Southern States of America. The epidermis or outer blackish bark is shaved off and the inner portion detached and ground or extracted. Although a very useful natural colouring matter, it has been considerably replaced by the cheaper Old Fustic. The colouring matter of quercitron consists of a tetraoxy flavanol termed quercetin.

```
                                OH
           O
HO/\/      \C—/  \OH
  |  |       ||  \__/
   \/\      
   OH \CO—COH
```

It is a polygenetic dyestuff, giving shades very similar to fisetin.

The dyestuff is present in the bark partly as a glucoside, quercitrin. The glucoside quercitrin, like the dyestuff quercetin, is also polygenetic.

Flavin is a commercial preparation of quercitron bark. It is obtained by extracting the bark with water at 103° C. Two brands are known: *yellow* flavin, which consists mainly of quercitrin, and *red* flavin, which is mainly quercetin. The red flavin is obtained by ammoniacal extraction of the bark.

Patent Bark is obtained by boiling quercitron bark in powder with three times its weight of 5 % sulphuric acid for about two hours; it is then filtered, washed and dried.

The preparations from quercitron bark are more powerfully tinctorial than the bark itself.

Weld consists of the plant *reseda luteola*; it contains mainly a textra-oxy-flavone **luteolin**.

```
                                OH
           O
HO/\/      \C—/  \OH
  |  |       ||  \__/
   \/\      
   OH \CO—CH
```

It is polygenetic, and is employed to a small degree for the dyeing of all fibres, but mainly silk.

Dyer's Broom (Dyer's greenweed) is found in Central and Southern Europe, in Russian Asia, and is frequent in the greater part of England. It contains a flavone luteolin, the dyestuff of weld, and a cumarone **genistein**, a feeble polygenetic dyestuff, which has the following constitution:

```
        /\  /  O  \
    HO |  |       CH—<  >OH
        \/\  \   /
       OH    CO
```

The yellows obtained from Fustic, Quercitron Bark and Weld are similar in their application and fastness. In the latter respect they are good, but not equal to the best acid dyes as regards light-fastness. Nevertheless, Fustic and preparations of Quercitron Bark were largely used even before the European War occasioned a shortage of the competing artificial products.

Indian Yellow (Piuri) is obtained by heating fresh urine of cows fed upon mango leaves. The yellow mass that separates is pressed in cloths and is then rolled into balls. In this form it is the magnesium salt of euxanthinic acid

```
 /\— O —/\
|  |—CO—|  |—O . CH . CHOH . CHOH . CH . CHOH
 \/      \/       |__________O_________|    |
 |                                          COOH
 OH
```

which on hydrolysis gives euxanthone

```
 /\— O —/\
|  |—CO—|  |OH
 \/      \/
 OH
```

It is mainly employed for colouring oil and water paints.

Cochineal consists of the dried insect *coccus cacti*, largely cultivated in Mexico. **Carminic acid**, the colouring matter of cochineal, probably has the empirical formula $C_{22}H_{22}O_{13}$. It crystallises from water or alcohol in red prisms; it darkens without melting at 130° C., and yields with alcoholic potash a mono- and a di-potassium salt. On distillation with zinc dust naphthalene is obtained. It was formerly much used on a tin mordant for the scarlet or "grain" dyeing of wool and silk, but its use is now declining, as although fast to light it cannot compete with acid scarlets in other respects. It is used also as a tint for microscopic work, and for the colouring of foodstuffs, etc.

Lac Dye is produced by an insect *coccus lacca* or *ficus*. Stick lac used for dyeing is deep red or brown and contains about 70 per cent. of resin and 10 per cent. of dye. The lac dye is obtained by extracting the stick lac with soda carbonate solution, evaporating the extract and moulding the concentrated extract into cakes. For dyeing purposes the lac dye is treated with dilute hydrochloric acid to remove mineral matter. It is polygenetic and is similar to cochineal, giving shades of good fastness to light but turned bluish by milling. Although faster to light than cochineal it has almost entirely gone out of use, while cochineal has remained in fairly extensive application. The main reason for this is the greater insolubility of lac dye, which can only be overcome by more laborious preparation for dyeing, e.g., grinding with hydrochloric or oxalic acid.

Archil, **Orchil** and **Cudbear** are prepared from certain lichens, *roccella tinctoria*, *lecanora tartarea* etc., by submitting them to the action of oxygen (air) in presence of ammonia, or the plant is boiled with lime and water, allowed to settle, ammonia added and exposed to the air until dyestuff no longer separates. Archil appears in commerce as a pasty mass, and a reddish powder, cudbear.

The dyestuff can be obtained from **Orcin** by the rapid action of hydrogen peroxide and ammonia, or slowly by the action of air and ammonia. The products vary according to the time of action of the reagents.

CH_3

HO OH Orcin

The colouring matter (**orceïn**) dyes best from a neutral bath, but it may be dyed from slightly acid or alkaline baths. It is used for the dyeing of crimson on wool and silk, with or without mordants. The dye is largely used in conjunction with indigo for the dyeing of wool, being used both as a bottoming and as a topping colour. Although rather fugitive according to modern standards, it is still used for crimson in silk dyeing.

Catechu or **cutch** is a valuable dyestuff obtained from the dark red heart wood of *Acacia*, *Areca*, and *Uncaria*, growing in India. It is largely used for obtaining various shades of brown, olive, drab, grey, and black, and is employed for dyeing both animal and vegetable fibres.

In practice catechu is now seldom used, as a product "prepared cutch," which is obtained by heating catechu with aluminium sulphate, possesses greater colouring power.

Cutch is remarkable for the brown colour produced on cotton by chroming, the colour being very fast to light, soap, alkali, acid, and also to bleaching liquor. Its use on wool is restricted owing to rendering the fibre harsh, but large quantities are used as a mordant in dyeing blacks on silk.

Catechin, one of the substances present in catechu, has the following constitution:

$$\begin{array}{l} CH_2O{-}O\diagdown \\ | \qquad\qquad\quad C_6H\,(-OH)(-CHOH-C_6H_3(OH)_2) \\ CH_2 \text{———} \\ \qquad\qquad\quad | \\ \qquad\qquad\quad OH \end{array}$$

Gambier catechu (or "**Gambier**"), in addition to catechin, contains a small amount of catechutannic acid, this substance being present in large quantities in the deep brown varieties of cutch. Catechutannic acid is said to be produced when catechin is heated to 110° C. It is a powerful tanning agent. Catechin is not precipitated by gelatin, and is not itself a tannin matter; it is, however, absorbed by hide, and there gradually passes into catechutannic acid.

Soluble Red Woods.—Brazil Wood, Peach Wood, Sapan Wood, and Lime Wood. These dyewoods are obtained from various species of *caesalpinia.* They have similar dyeing properties, and on account of the fugitive character of the colours they yield they are only used to a limited extent.

The woods contain the leuco dyestuff **brasilin.**

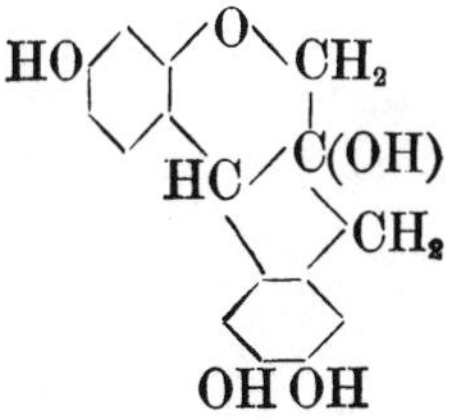

HO O CH_2
C(OH)
HC
CH_2
OH OH

This oxidises in aqueous solution by exposure to air to the dyestuff **Brasileïn**.

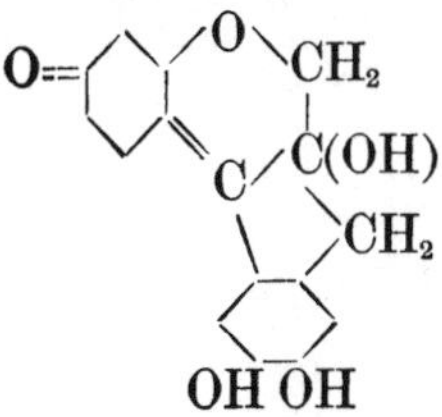

O= O CH_2 or HO O CH_2
C(OH) C(OH)
C C
CH_2 CH_2
OH OH OH O

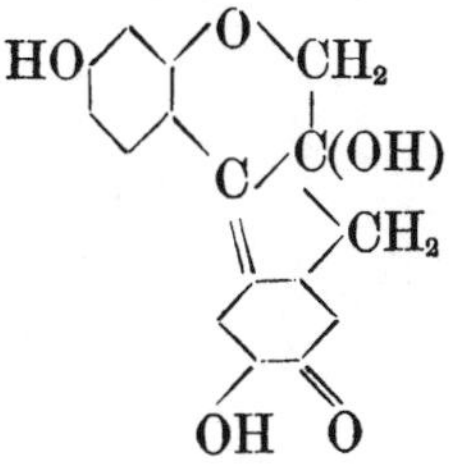

Insoluble Red Woods.—**Camwood**, **Barwood**, and **Sanderswood** or **Sandal Wood.** The wood contains the dyestuff termed **santalin**, present to about 17 °/₀ in sanderswood and 23 °/₀ in barwood. These dyewoods are obtained from certain species of *pterocarpus* and *baphia*.

They are principally used in wood-dyeing, in conjunction with other dyestuffs. In some combinations, e.g., with logwood, their inferior fastness does not reveal itself as might be imagined from separate tests. The dyestuff is polygenetic.

Logwood or **Campeachywood** consists of the heart wood of *haematoxylon campechianum* growing chiefly in Central America, Mexico, etc. The best sorts come from Mexico, Haiti, San Domingo, Honduras, Cuba, Jamaica, and Guadeloupe. There is also on the market an extract of the wood of variable strength. The principal use of logwood is for the production of blacks upon both animal and vegetable fibres, and it is often used for this purpose along with Fustic. It is a polygenetic dyestuff and gives, with an aluminium mordant, a blue colour; chrome, blue-black; iron, black; copper, green-black; and tin, violet.

The wood contains the leuco compound **haematoxylin**, and is therefore similar in constitution to brasilin.

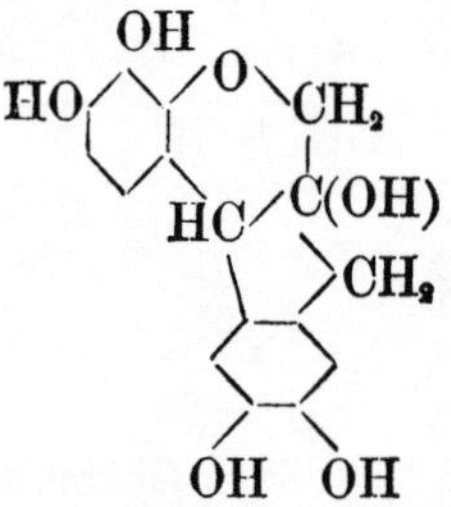

OH
HO O CH_2
HC C(OH)
CH_2
OH OH

By oxidation of haematoxylin, generally in ammoniacal solution by exposure to the air, **haematein** is produced, which probably has the following constitution:

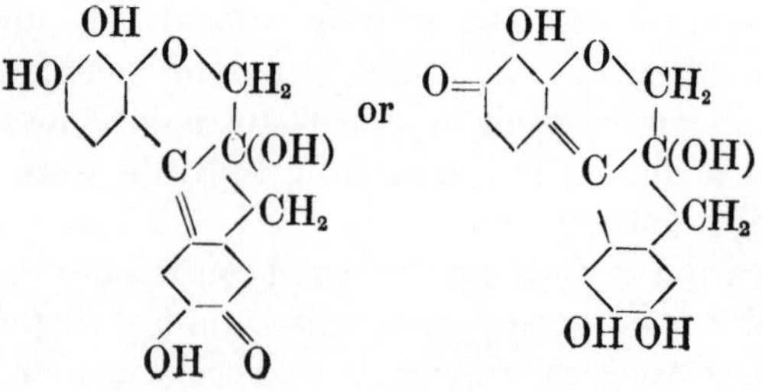

OH
HO O CH_2
C C(OH)
CH_2
OH O

or

OH
O= O CH_2
C C(OH)
CH_2
OH OH

Haemateïn is converted into a colourless body by sulphurous acid or sodium bisulphite solution, readily soluble in water. No reduction, however, appears to occur, as on addition of an acid or on boiling, haemateïn is precipitated; also with zinc and acid, with stannous chloride and caustic soda, the solution is decolourised, but on standing the solution is again restored to its former tint.

Logwood used to be "aged." This process consists in piling the "chips" or "rasps" in a moist condition in a large airy chamber. The mass oxidises and the yellow wood turns dark reddish brown. At the present time this method is seldom used, as formerly iron was chiefly used as mordant, whereas bichromate is the mordant mainly used now for blacks, and the yellow chrome mordant oxidises the haematoxylin to haemateïn.

Logwood extract is still oxidised to obtain haemateïn, the oxidation being mainly done by means of sodium nitrite.

Commercial **Haemateïn crystals** are among the best forms of logwood extract, and are usually much stronger in colouring power than solid logwood extract.

Woad is a dark clay-like product made from the leaves of the woad plant, *isatis tinctoria.* The plant is a biennial, indigenous to Europe, and has long been used for dyeing blue. The ancient Britons stained their skin blue with woad in time of war and in connection with certain religious observances. It was largely cultivated before the introduction of indigo from India. Woad is still grown in Lincolnshire and Huntingdon. The leaves are subjected to a fermentation process to prepare it for the market. Woad, although previously used for dyeing blue, is now only employed as a fermenting agent in the indigo-vat as used by the wool-dyer. The vat so prepared is therefore termed the "woad-vat." Woad was considered to contain the same colouring matter as is present in the *indigoferae*. It is now known to be a distinct substance, which in most of its reactions resembles indoxylic acid.

(The student seeking further information on natural dyestuffs is recommended to consult the various articles by A. G. Perkin in Thorpe's *Dictionary of Applied Chemistry*, also *Die natürlichen Farbstoffe*, Rupe.) The slight commercial importance of natural dyes as a class does not warrant further space being assigned to them in this book. All those dealt with above are still more or less in actual use, but only logwood, indigo, cutch and fustic are now of considerable importance in normal times. The famine in synthetic dyestuffs experienced by most nations except Germany during the European war, caused a transient revival of interest in the use of natural products for dyeing. This, however, is not likely to be repeated, as in most cases steps have been taken to prevent any future recurrence of shortage from the same cause.

APPENDIX

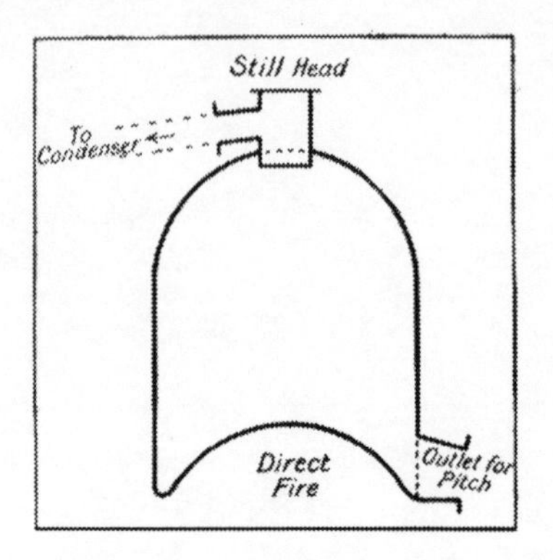

Fig. I. Ordinary Tar Still.

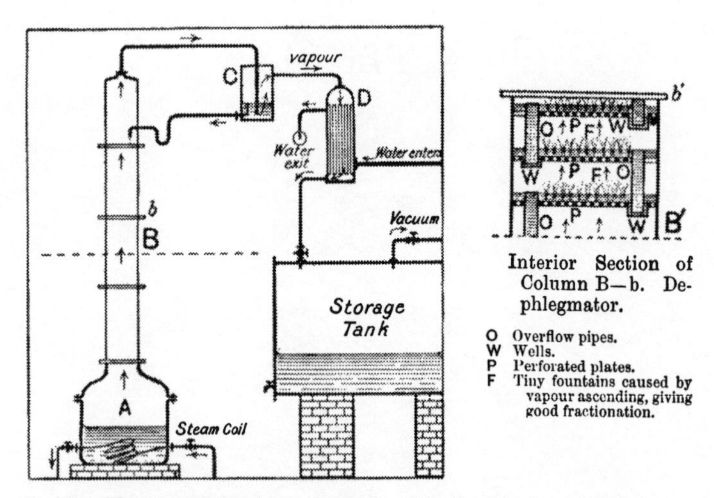

Interior Section of Column B—b. Dephlegmator.

O Overflow pipes.
W Wells.
P Perforated plates.
F Tiny fountains caused by vapour ascending, giving good fractionation.

Fig. II. Plant for Distillation of Coal Tar Oils, Rectification of Oils, etc.

A Still, steam-heated.
B Savalle column, still-head.
C Air-cooler or condensator.
D Condenser proper (water tubes).

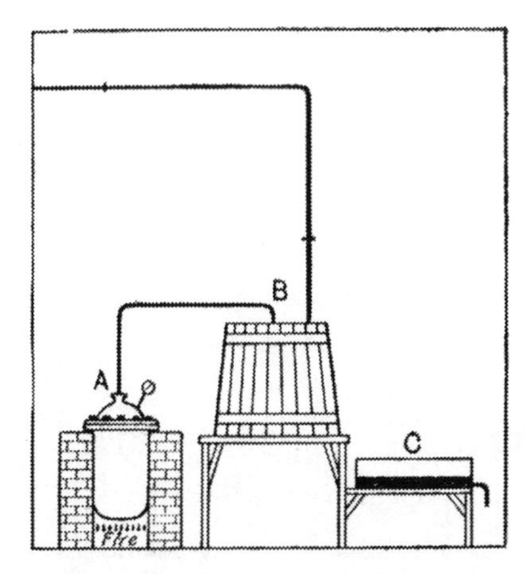

FIG. III. Plant for Manufacture of Diphenylamine Products.

A Autoclave.
B Vat for washing, dissolving out impurities.
C Table filter.

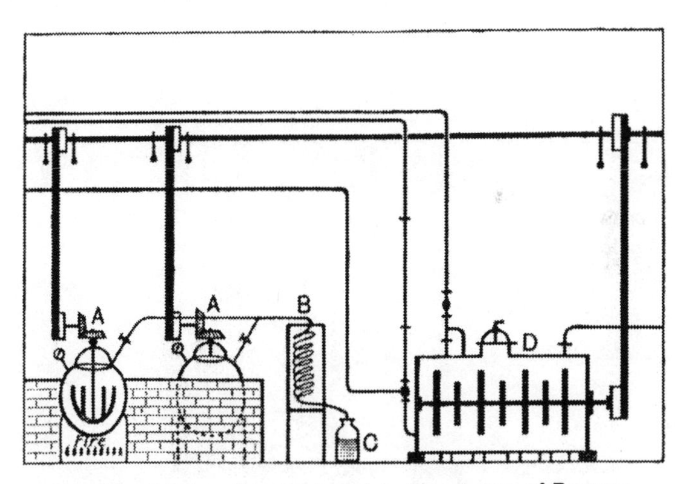

FIG. IV. Sulphonation Plant used in Manufacture of Benzene Sulphonates, etc.

A Autoclave with Stirrer and Gauge.
B Condenser for Excess Hydrocarbon.
C Receiver.
D Agitating vessel for neutralisation, "liming-out," etc.

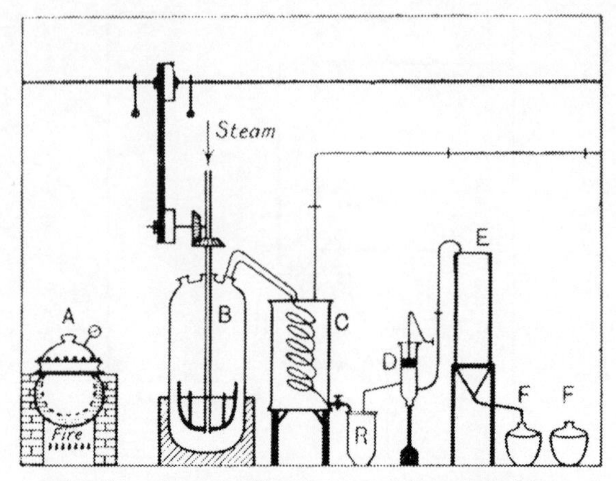

Fig. V. Plant for Manufacture of Dimethylaniline.

A Autoclave in oil-bath.
B Still (stirrer with hollow spindle for steam).
C Condenser.
R Receiver.
D Feed-pump.
E Separating vessel.
F Carboy.

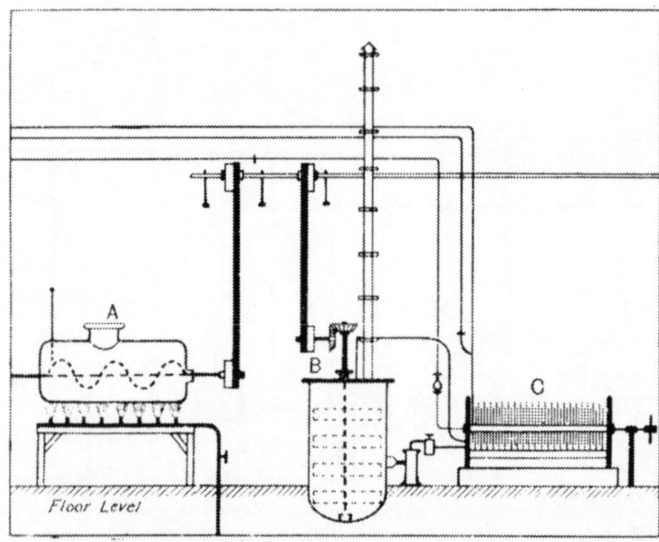

Fig. VI. Plant for Manufacture of Naphthol products from Sulphonic Acids, e.g. Amidonaphthol sulphonic acid Y (2 . 8 . 6) from β-naphthylamine disulphonic acid G by heating under pressure with alkali.

A Pressure vessel for alkali melt.
B Vessel for neutralisation of alkali melt (with stirrer).
C Filter press.

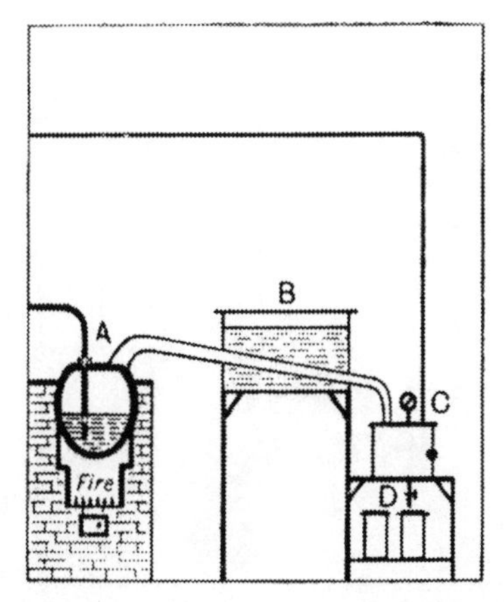

Fig. VII. Plant for Rapid Distillation and Purification. Used in purification of Resorcine.

A Retort-still.
B Condenser.
C Receiver.
D Moulds.

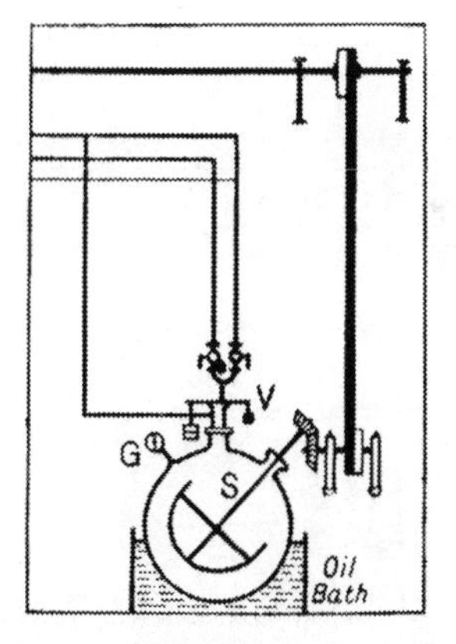

Fig. VIII. Plant for Alkylating, etc. under Pressure.

G Pressure gauge. V Safety valve. S Stirrer.

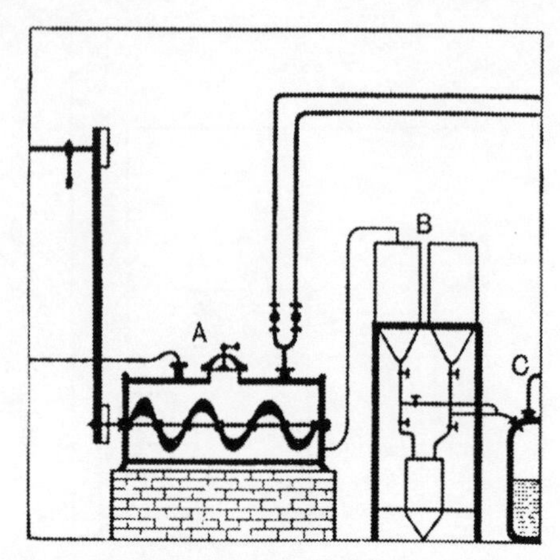

Fig. IX. Plant for Solvent Extraction and Separation, e.g. used of extraction of Resorcine with Amyl Alcohol and Ether.

A Extraction vessel with spiral agitator.
B Separating vessels.
C Still for recovery of solvent.

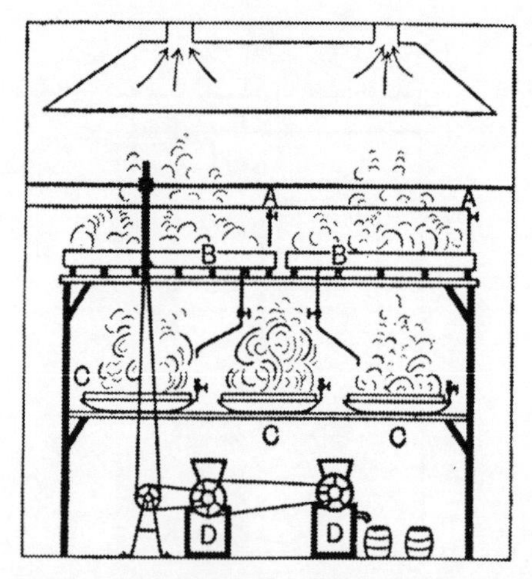

Fig. X. Evaporating and Grinding Plant used in Manufacture of Intermediate Products and Dyestuffs, Dyewood Extracts, etc

A Supply pipes.
B Reservoir heaters.
C Evaporating pans (steam-jacketed).
D Grinding mills.

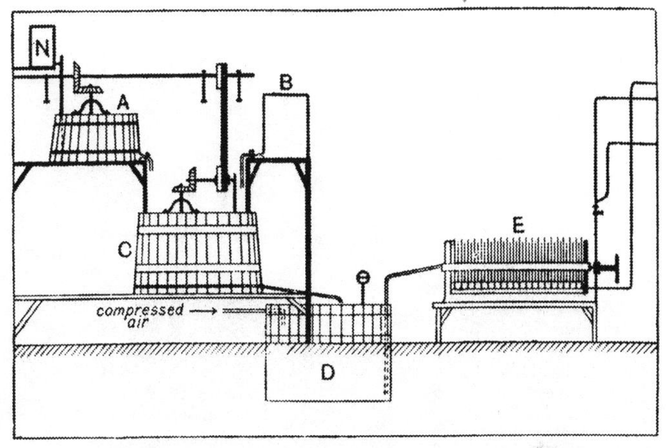

Fig. XI. Plant for Manufacture of an Azo Dyestuff.

A Vat for diazotising the amine component (with paddle agitator).
B Vessel for dissolving second component (phenol or amine).
C Coupling vat (with agitator).
D Monteju (receiving reservoir).
E Filter press.
N Vessel for sodium nitrite.

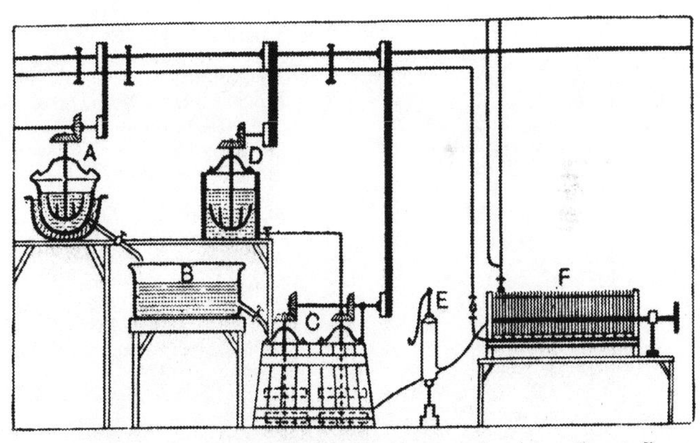

Fig. XII. Special Plant for Manufacture of an Azo Dyestuff.

A Autoclave for sulphonation, or dissolving, or diazotisation under pressure.
B Receiver from A for neutralisation, dilution, etc. if required.
C Coupling vat (tandem paddle stirrers).
D Vessel for dissolving or diazotising second component.
E Pump feeding press.
F Filter press.

INDEX

The chief reference is given in heavy type.

Printed in the United States
By Bookmasters